Microscale Chemistry

Compiled by John Skinner

Teacher Fellow
The Royal Society of Chemistry
1994–1995

D1580676

Microscale Chemistry

Compiled by John Skinner

Edited by John Johnston and Colin Osborne

Designed by Imogen Bertin

Published by the Education Division, The Royal Society of Chemistry

Printed by The Royal Society of Chemistry

For further information on other educational activities undertaken by the Royal Society of Chemistry write to:

The Education Department
The Royal Society of Chemistry
Burlington House
Piccadilly
London W1V OBN

ISBN 1 870343 49 2

British Library Cataloguing in Data.
A catalogue for this book is available from the British Library.

THE ROYAL
SOCIETY OF
CHEMISTRY

Contents

THE ROYAL
SOCIETY OF
CHEMISTRY

Foreword

The use of microscale chemistry is gaining momentum worldwide and is now an integral part of courses in North America, France and South Africa as well as in the UK.

The advantages of microscale chemistry are evident in terms of safety and convenience. Microscale experiments also require students to rethink their approach to experimental technique and encourages increased accuracy and skill in carrying out procedures.

Microscale chemistry in the UK is currently mainly confined to undergraduate teaching in universities. However, The Royal Society of Chemistry believes that this publication will provide the basis to help establish microscale chemistry at the secondary level.

Professor E W Abel DSc CChem FRSC
President, The Royal Society of Chemistry

THE ROYAL
SOCIETY OF
CHEMISTRY

Introduction to microscale chemistry

Devising microscale organic chemistry experiments has been a recent initiative in UK science education.

This development has been mostly at undergraduate level through the efforts of Dr Stephen Breuer at the University of Lancaster. A book of experiments has been published as well as a special set of smallscale glassware which is now commercially available. The techniques are gradually being incorporated into undergraduate courses at other universities. A survey of students' views at Lancaster University on the use of microscale experiments has found a generally positive response. In addition, it is now regarded as 'normal' to do microscale organic chemistry.

At secondary level, organic chemistry experiments based on special smallscale Quickfit glassware have received a rather mixed response from teachers. In schools and colleges most organic chemistry is done at the post-16 level, when a number of undergraduate experiments might be appropriate. However, there are difficulties in simply using undergraduate experiments in post-16 courses.

First, many experiments require specialised glassware and although it might be possible for schools to purchase one or two kits it is unlikely that they would be prepared to equip a whole class. In addition, breakages are costly to replace.

Secondly, the time available to undergraduates and post-16 students to do an experiment differs markedly. Many organic preparations are slow and are difficult to fit into a post-16 session. For instance, refluxing for 1 hour may be appropriate at university but it is too long for most school or college timetables.

Thirdly, in devising organic experiments for schools we need to look for characteristics in compounds which can be identified readily using simple methods. Since most schools and colleges do not have access to the spectroscopic equipment available to undergraduates – tests based on colour, melting point, the presence of double bonds, precipitate formation, and in some cases for odour – eg esters – are required.

Despite these constraints, there is a whole range of interesting microscale experiments that can be done in each of the three areas of organic chemistry: tests for functional groups, their interconversion, and synthesis. It is helpful that many (but not all) of the organic compounds used in schools and colleges such as alkanes, alkenes, alcohols, phenols, esters and carboxylic acids, are compatible with the smallscale plastic apparatus used in general chemistry experiments – ie well-plates, pipettes and petri dishes. In addition, many organic experiments do not require heating and of those that do, a warm water bath may be sufficient.

Two of the main features of the experiments in this book are the reduction in use of hazardous reagents and the ease of disposing of waste products. For example, tests for unsaturation using bromine are very widely used in pre-16 and post-16 organic chemistry and usually require bottles of bromine or bromine dissolved in halogenated solvents. In microscale unsaturation tests, the bromine is actually generated as required immediately before the test and very little, if any, is left for disposal. If tests like these gain wide acceptance, then there will be a major reduction in using hazardous reagents in schools.

Inorganic and general chemistry can also benefit from microscale techniques. Experiments can be transferred from conventional test-tube activities and can be done in well-plates and on clear plastic sheets. Specific techniques – eg titrations – can also be modified successfully for use on microscale.

THE ROYAL
SOCIETY OF
CHEMISTRY

Background

John Skinner

Microscale chemistry is chemistry carried out on a reduced scale using small quantities of chemicals and often (but not always) simple equipment. This book contains a wide range of microscale chemistry experiments designed for use in schools and colleges in the UK. In the US the term smallscale chemistry is often preferred, especially at secondary level. There, the term microscale is normally used for organic experiments done in specialised glassware. In this book, however, the word microscale is used throughout.

Microscale experiments have several advantages:

▼   the small quantities of chemicals and simple equipment reduces material costs;

▼   the disposal of chemicals after the experiment is easier;

▼   safety hazards are often reduced and many experiments can be done quickly; and

▼   frequently, plastic apparatus can be used rather than glassware so breakages are minimised.

Microscale chemistry is not just about doing conventional experiments on a reduced scale. Often experiments can be done in new ways such as precipitation reactions in drops of liquids. Chemical reactions often proceed in small volumes of solution rather than the much larger volumes in test-tubes and beakers, and it is often possible to make observations at microscale that are not possible at normal scale. Such experiments will teach students the importance of careful observation – a skill that is vitally important in any scientific endeavour. By minimising hazards microscale chemistry opens up the possibility of using chemicals that are too hazardous to contemplate on a larger scale thereby increasing students' experience of practical chemistry. On a general level, by minimising waste, microscale chemistry encourages students to use chemicals responsibly, an issue very much in tune with current environmental concerns.

The microscale experiments described in this book are not intended to replace conventional experiments. Some may substitute for those in current practice for safety reasons – *eg* in tests for unsaturation with bromine; of cost – *eg* using microscale titrations with silver nitrate solution; or speed – *eg* making esters. Sometimes the experiments may be used to complement existing methods by allowing students to perform these experiments either alongside existing methods at appropriate points in a course or as quick but useful revision exercises at the end of a topic.

Because many of the experiments are novel and unusual, many students (and some teachers) may take time to get used to some of the techniques. For example, using a plastic pipette requires a steady hand and the application of the correct amount of pressure to the bulb. Nevertheless if a mistake is made the drops may be quickly mopped up with a tissue and very little chemical will have been wasted or time lost. With practice students should find the techniques are easy to use. The emphasis throughout this book is on maximising the opportunities for careful observation and interpretation. Therefore the practical parts of these experiments are often over very quickly (in only a few minutes) in contrast to traditional experiments

THE ROYAL
SOCIETY OF
CHEMISTRY

which might last a whole lesson. However, this is not to say that experimental technique is unimportant.

You are encouraged to sample the experiments and to assimilate gradually some of them into your courses where appropriate. This book does not claim to cover all the experiments that could be done in microscale and you may wish to devise some of your own, by developing some of the ideas and techniques shown here.

Current trends indicate that, with the likelihood of further environmental legislation, the need for microscale chemistry techniques and experiments in schools and colleges is likely to grow. This book should serve as a guide in this process.

THE ROYAL
SOCIETY OF
CHEMISTRY

Historical developments

John Skinner

The benefits of doing smallscale experiments were realised in the late 1970s and early 1980s mainly in the US. The impetus for change was part environmental and part economic. Increasing public concern over the potential damage to the environment due to chemicals, particularly the need to improve air quality inside laboratories and reduce the cost of chemicals, were the main driving forces behind the so-called 'downsizing' of experiments in organic chemistry.

One of the pioneers of downsizing was Stephen Thompson at Colorado University, who developed many new smallscale experiments, published in *Chemtrek*. Several groups formed in the north-eastern US to further develop microscale techniques and several publications on organic and general chemistry were produced during the late 1980s especially those by Professors Mayo, Pike, Butcher and Trumper.

In 1993 the National Microscale Chemistry Centre (NMC[2]) was established at Merrimack College in Massachusetts. The aim of the Centre, under Director Dr Mono Singh, is to promote the use of microscale chemistry as a way of preventing pollution by eliminating waste at the source. The Centre holds courses and workshops for science teachers. Recent legislation in the US has provided a strong impetus for introducing microscale techniques and experiments; in many states it is illegal to pour any chemicals down the sink!

Recognising the increased interest amongst teachers for microscale chemistry the American Chemical Society's *Journal of Chemical Education* began publishing a section entitled *The microscale laboratory* in 1989 to report experiments ranging from lower secondary school to advanced undergraduate level.

Outside the US, several groups in Canada have promoted microscale chemistry especially at secondary education level. Over the past few years Professor Geoffrey Rayner-Canham and Dr Alan Slater have been among those reporting key experiments in *Chem 13 News* published by the University of Waterloo, Ontario.

In the UK, progress at undergraduate level has been made over the past few years largely through the pioneering efforts of Dr Stephen Breuer at the University of Lancaster who has produced a manual of microscale organic chemistry experiments that has been used in chemistry courses at the university. These experiments are now being used in other universities.

Although microscale chemistry is already being practised in some areas of secondary education there is an overall lack of awareness of the techniques and methods of microscale chemistry, leaving a gap which this book aims to fill. In addition many publications on microscale chemistry, which might have encouraged teachers to adopt microscale chemistry in their courses, are difficult to obtain in this country. A summary of some current techniques and examples of experiments are, however, given by Professor Rayner-Canham in *Education in Chemistry*, May 1994.

THE ROYAL
SOCIETY OF
CHEMISTRY

Other developments

For a number of years individuals around the world have been looking into the design of low cost equipment for practical work in chemistry. This reflects the awareness of the high costs of laboratory equipment and consumables. More recently safety and waste disposal concerns have prompted the search for improvements in chemical practice. In South Africa our group has developed an approach which draws upon the efforts of individuals in different countries.

The outcome of our collaboration with colleagues throughout the world has led to the development of a microscale chemistry kit and a series of worksheets on a wide range of practical activities. Trialling in South Africa was successful and the majority of teachers and lecturers were extremely supportive of the initiative in terms of ease of procedure and the time taken to complete experiments.

Our local experiences encouraged us to trial our system in other countries and this has now been done in the US, UK, Australia, France, Finland, Egypt, the Philippines, Ghana, Kenya, Zambia, Mozambique and Malawi. There has been great interest, and even language barriers do not dampen enthusiasm. Senior science educators at a conference at the University of Montpellier saw in the system the potential to change the face of chemistry teaching in their schools. UNESCO officials reached a similar conclusion and proposed the implementation of a global programme. This reaction has given us hope that our system can help to change some of the approaches to practical chemistry.

Professor J.D. Bradley
University of Witwatersrand
Johannesburg

THE ROYAL
SOCIETY OF
CHEMISTRY

List of experiments

THE ROYAL
SOCIETY OF
CHEMISTRY

THE ROYAL
SOCIETY OF
CHEMISTRY

Safety

We believe that the experiments in this book can be done safely, but it is the responsibility of the teacher to make the final decision depending on the circumstances at the time. Teachers must ensure that they follow the safety guidelines set down by their employers. A risk assessment must be completed for any experiment that is done.

Aminobenzene is toxic by inhalation, in contact with the skin and if swallowed.

Ammonia solution causes burns and gives off ammonia vapour which irritates the eyes, lungs and respiratory system.

880 ammonia causes burns and gives off a vapour that irritates the eyes, lungs and respiratory system.

Barium compounds are harmful by inhalation and if swallowed

Bleach is an irritant.

Bromine vapour is an irritant and is very toxic if inhaled.

Calcium carbide liberates flammable ethyne gas on contact with water.

Chlorine is toxic. It is harmful to the eyes, lungs and respiratory tract.

Cobalt compounds are harmful if swallowed.

Copper compounds are harmful to the skin, eyes and lungs, and are toxic when ingested.

Cyclohexane is flammable and has a harmful vapour. It is irritating to skin and eyes.

Cyclohexene has a harmful vapour that irritates the eyes, skin and respiratory system.

2,4-Dinitrophenylhydrazine is irritating to the skin and eyes. It is harmful if inhaled, ingested or absorbed through the skin.

Ethanoic anhydride is flammable and causes burns.

Ethanal is highly flammable and is irritating to the eyes and respiratory system.

Ethanol is flammable.

Ethoxyethane is extremely flammable and may form explosive peroxides.

Ethyne is extremely flammable.

Hexane is flammable.

THE ROYAL
SOCIETY OF
CHEMISTRY

Hydrochloric acid can cause burns. It gives off an irritating vapour that can damage the eyes and lungs.

Concentrated hydrochloric acid causes burns. It gives off an irritating vapour which can damage the eyes and lungs.

Hydrogen peroxide is corrosive, irritating to the eyes and skin, and is a powerful oxidising agent.

Hydrogen sulphide is very toxic by inhalation.

Hydroxybenzene (Phenol) is toxic by ingestion and skin absorption. It can cause severe burns. Take care when removing phenol from the bottle because the solid crystals can be hard to break up. Wear rubber gloves and a face mask.

2-Hydroxybenzoic acid (salicylic acid) is harmful by ingestion and is irritating to the skin and eyes.

Iodine is harmful by skin contact and gives off a toxic vapour that is dangerous to the eyes.

Iron(III) nitrate is an oxidising agent that may assist fire.

Lead compounds are toxic.

Lead nitrate is an oxidising agent that assists fire.

Methanol is highly flammable and is toxic by inhalation and if swallowed.

Molybdenum compounds are irritants and nearly all are harmful by inhalation or ingestion.

2-Methylpropan-2-ol gives off a vapour that is irritating to the skin and eyes.

2-Naphthol is harmful by inhalation and if swallowed.

Nickel salt solutions are irritating to the eyes and are assumed to be poisonous if swallowed.

Nitrates are powerful oxidising agents and assist fire.

Nitric acid causes severe burns and is an oxidising agent that may assist fire.

Concentrated nitric acid causes burns. Its vapour can damage eyes and lungs.

Nitrogen dioxide is toxic and irritates the respiratory system.

Phosphoric acid is irritating to the eyes and causes burns.

Potassium bromate(V) is a strong oxidising agent and is toxic by ingestion.

Potassium dichromate is a powerful oxidising agent. It can cause ulcers on contact with the skin and is a suspected carcinogen.

Potassium iodate is an oxidising agent and can assist fire.

Potassium permanganate is a skin irritant and is harmful if swallowed. It can cause fire on contact with combustible materials.

Propan-1-ol is highly flammable.

Propan-2-ol is highly flammable.

Propanone is flammable.

Silver nitrate causes burns. It is an oxidising agent and may assist fire.

Sodium fluoride is toxic by inhalation, in contact with skin and if swallowed.

Sodium hydroxide can cause severe burns to the skin and is dangerous to the eyes.

Sulphur dioxide is an acidic and choking gas. It is toxic by inhalation and can affect asthma sufferers adversely.

Sulphuric acid is corrosive and causes burns.

Concentrated sulphuric acid is a strong acid, a powerful oxidant and dehydrating agent. Its reaction with water is highly exothermic.

2,4,6-Trichlorohydroxybenzene is harmful if swallowed and is irritating to the eyes and skin.

Tungsten compounds are harmful by inhalation, ingestion and skin absorption.

Urea is irritating to the skin eyes and respiratory system.

Vanadium (V) salts are toxic by ingestion or by inhalation of dust. They irritate the skin.

Zinc chloride causes burns.

THE ROYAL
SOCIETY OF
CHEMISTRY

Acknowledgements

I thank the many individuals and organisations that have helped with this project – by trialling experiments, giving advice and offering suggestions. Without them this project would not have been possible.

The following people contributed ideas and/or trialled experiments:

D. M. Arnott	King's School, Canterbury
Linda Ashby	Abbey School, Kent
John Broadbent	Leek High School, Staffordshire
Mike Broderick	Brynteg Comprehensive School, Bridgend
Andrew Browning	Canford School, Dorset
Jane Carrington-Porter	Goodyer's End Middle School, Nuneaton
T. Challenger	Brighton College of Technology, Brighton
May Chan	Buckingham College, Harrow
Neil Coombes	Ysgol Friars, North Wales
Elaine Coop	Colyton Grammar School, Devon
Noel Dickson	Watford Boys' Grammar School, Watford
Justin Dillon	King's College, London
Ben Faust	Loughborough Grammar School, Loughborough
Tim Fearn	Trent College, Nottingham
J. Gauld	St.Wilfrid's C of E High School, Blackburn
David George	Cranbrook School, Kent
Anne-Marie Gielty	Boundstone Community College, West Sussex
David Glasby	Craigmount High School, Edinburgh
Nicola Hampstead	St.Catherine's School, Guildford
R. J. Hill	Benenden School, Kent
J. M. Hillier	Manchester High School for Girls, Manchester
Norman Hooper	Sexey's School, Somerset
Noel Jackson	Belmont Comprehensive School, Durham
Pat Jackson	St. Mary's Hall, Brighton
Alan Jobber	Sparrow Farm Junior School, Epsom
Jacky Jordan	Ernest Bevin School, London,
Gavin Lazaro	Burnham Grammar School, Buckinghamshire
Ted Lister	Trinity School, Leamington Spa
P. McKerchar	Hurstpierpoint College, Sussex
Judy Machin	Gumley House School, Middlesex
Ross Macpherson	Saltash Community School,Cornwall
Tim Meunier	Dragon School, Oxford
Sally-Jane Morris	Cotswold School, Bourton-on-the-Water
Susan Munday	Malvern Girls' College, Malvern
Valerie Myers	Blackheath High School, London
Jane Patching	Wickersley Comprehensive School, Rotherham
David Peacock	Great Marlow School, Buckinghamshire
Claire Penfold	Penrice School, Cornwall
Paul Pinder	De Ferrers High School, Burton upon Trent
Steve Rhode	St. Paul's School, Haywards Heath
Ray Sabine	Sutton Valence School, Kent
D. Sanders	Hastingsbury Upper School, Bedford
Richard Sanderson	Tiffin School, Kingston upon Thames
E. A. Skinner	Laindon School, Basildon
John Starling	Sackville School, East Sussex
Roger Suffolk	St. Andrew's C of E High School for Boys, Worthing

THE ROYAL
SOCIETY OF
CHEMISTRY

Stephen Taylor	Bishop's Stopford School, Enfield
Sandra Taylor	Banbury School, Banbury
Jayne Wainwright	Colyton Grammar School, Devon
Dave Waistnidge	King Edward VI College, Devon
Rod Watson	King's College, London
John Way	Parmiter's School, Watford
A.R. Williams	Sydenham High School for Girls, London
I. B. C. Wood	Hastingsbury Upper School, Bedford
P. H. Wright	Ilford County High School, Ilford

Special thanks is due to David Moore of St. Edward's School, Oxford and John Davies of Hipperholme and Lightcliffe High School, Halifax.

I also thank Professor Bob Silberman of the State University of New York for his advice, and for many interesting ideas that have been developed in this project; Dr Stephen Breuer of the University of Lancaster and Dr Jim Hanson of the University of Sussex for assisting with organic chemistry experiments; Dr Mono Singh and Professor Ronald Pike of the National Microscale Chemistry Centre, Massachusetts, US for interesting discussions; staff at the Royal Institution (RI) who supported the project, including Professor Peter Day who kindly provided laboratory space and David Madill and Mike Sheehy who gave invaluable technical support; Anthony Graham, Cameron Kepert, Martin Morris and Chris Nuttall, with whom I shared a laboratory at the RI, and who helped with the trialling of experiments; and my son Robert who provided the illustration for experiment 20.

Finally, I thank the staff of the education department at The Royal Society of Chemistry, especially Neville Reed and John Johnston, for their advice, encouragement and unfailing support throughout the project.

The Royal Society of Chemistry thanks the Royal Institution for providing laboratory space for the duration of this project, and the Head and Governors of Tiffin School for Girls for seconding John Skinner to the Society's Education Department to carry out this project.

THE ROYAL
SOCIETY OF
CHEMISTRY

Bibliography

Current books and journals on microscale chemistry are mainly published in North America. The following textbooks have provided useful resources.

Books

Stephen Thompson, *Chemtrek – small-scale experiments for general chemistry* New Jersey: Prentice-Hall, 1990.

Stephen W. Breuer, *Microscale practical organic chemistry.* Lancaster: Lancaster University, 1991.

Judith C. Foster, Ronald M. Pike and Zvi Szafran, *Microscale general chemistry laboratory.* New York: John Wiley, 1993.

C. B. Allen, S. C. Bunce and J. W. Zubrick, *Annotated list of laboratory experiments in chemistry from the Journal of Chemical Education 1957–1992.* Washington DC: A C S, 1993.

Alan Slater and Geoffrey Rayner-Canham, *Microscale chemistry – laboratory manual.* Don Mills, Ontario: Addison-Wesley, 1994.

Dana W. Mayo, Ronald. M. Pike and Peter K. Trumper, *Microscale organic laboratory.* New York: John Wiley, 1994.

Edward L. Waterman and Stephen Thompson, *Small-scale chemistry – laboratory manual.* Don Mills, Ontario: Addison-Wesley, 1995.

Mono M. Singh, Ronald. M. Pike and Zvi Szafran, *Microscale and selected macroscale experiments for general and advanced general chemistry.* New York: John Wiley, 1995.

Journals

No journals are dedicated exclusively to microscale chemistry. The most useful and readily available publications in this field are:

Journal of Chemical Education. Since 1989 this journal has included a section called *The Microscale laboratory* describing experiments ranging from secondary level to third year undergraduate level.

Chem 13 News. This is produced monthly in Canada by the department of chemistry at the University of Waterloo and contains occasional articles on microscale chemistry especially for secondary education. (This journal is included in the Royal Society of Chemistry's *Schools Publications Service.*)

An article by Professor Geoffrey Rayner-Canham describing some of the smallscale apparatus used for general chemistry experiments appears in *Education in Chemistry*, 68, 1994 along with a list of references.

In November 1994 the American Chemical Society published a special issue of *Chemunity News* (vol 4, no 3) entirely devoted to smallscale chemistry.

J.D. Bradley *et al, J. Chem. Ed.,* 1997, submitted for publication.

J.D. Bradley *et al, Proc. 14th ICCE,* Brisbane, 1996.

J.D. Bradley *et al, Proc. CIFFERSE* Seminar, Montpellier, 1997, in press.

J.D. Bradley *et al, Chem. Proc. SA,* 1997, submitted for publication.

THE ROYAL
SOCIETY OF
CHEMISTRY

Apparatus and techniques for microscale chemistry

Teacher's Information Sheet

This section provides information about some of the apparatus, solutions and techniques that are needed in the microscale chemistry experiments described in this book.

Apparatus

1. Transparent plastic sheet (OHP sheet)

These sheets are widely used in schools and colleges and are used to overlie the student worksheets in several experiments. Students add drops of solutions onto the sheets to do the reactions. The sheets are reasonably resilient and may be wiped clean with household tissues and re-used many times. However, they could be attacked by strong acids and they are stained by iodine solution if left in contact for more than a few minutes.

Alternatively, clear plastic wallets (A4 size) can be used or the student worksheets could be laminated. Again the wallets or laminated sheets can be wiped clean after use.

Whichever type is used, aqueous solutions form nicely-defined drops on the surface which enable chemical reactions to be conveniently carried out. A discussion of the shape of the drops can provide students with interesting insights into the effects of hydrogen bonding on surface tension.

2. Plastic pipettes

These very versatile pieces of plastic apparatus may be used for storing solutions and dispensing drops of solution during experiments. By cutting and re-shaping them it is possible to make scoops and spatulas, filter funnels, mini reaction vessels, and build electrolysis apparatus. The two most useful forms of pipette are:

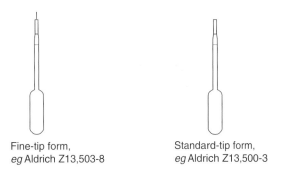

Fine-tip form,
eg Aldrich Z13,503-8

Standard-tip form,
eg Aldrich Z13,500-3

These plastic pipettes have many uses. Two of the most useful are preparing a shortened pipette for storing solutions and making a scoop.

Preparing a shortened pipette

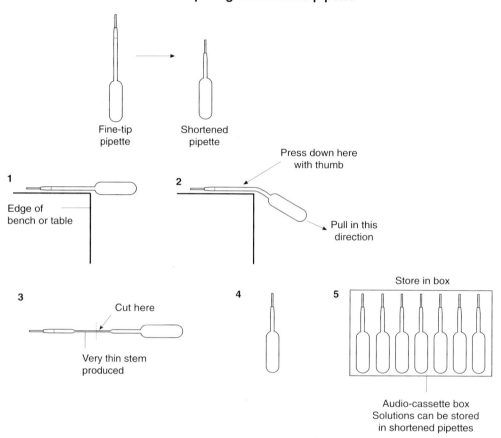

Fine-tip
pipette

Shortened
pipette

1

Edge of
bench or table

2

Press down here
with thumb

Pull in this
direction

3

Cut here

Very thin stem
produced

4

5

Store in box

Audio-cassette box
Solutions can be stored
in shortened pipettes

Making a scoop

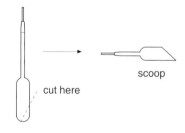

scoop

cut here

These pipettes are available from:
Aldrich Chemical Company Ltd,
The Old Brickyard,
New Road,
Gillingham,
Dorset SP8 4JL.
Tel: 01747 822211
Fax: 01747 823779

THE ROYAL
SOCIETY OF
CHEMISTRY

Plastic petri dishes

These are used as containers in which test gases can be generated. Examples are the 4.5 cm and 9 cm diameter sizes supplied by Philip Harris (ref: Y36340). For more information contact:

Philip Harris Education,
Lynn Lane,
Shenstone,
Lichfield,
Staffordshire WS14 0EE
Tel: 01543 480077
Fax: 01543 480068

Well-plates

Clear plastic well-plates are sometimes listed in catalogues under culture apparatus. The most useful version is the 24-well plate which consists of an arrangement of 6 x 4 cylindrical wells each well having a capacity of *ca* 3 cm³.

eg Sigma M 9655

These well-plates are used for organic chemistry experiments, equilibria, rates of reaction and colorimetry experiments. They are compatible with many organic liquids.

Other apparatus

Other useful pieces of equipment are audio-cassette boxes for storing shortened plastic pipettes, plastic boxes for 35 mm film, plastic display boxes for pens and screw-top sample bottles. Temperature strips can be ordered from Sandra Bell, Education Department, The Royal Society of Chemistry, Burlington House, Piccadilly, London W1V 0BN.

Element solutions

Barium

Barium nitrate solution (0.2 mol dm^{-3}) – dissolve 5.2 g Ba(NO$_3$)$_2$ in 100 cm³ of deionised water.

Calcium

Calcium nitrate solution (0.5 mol dm^{-3}) – dissolve 11.8 g of Ca(NO$_3$)$_2$.4H$_2$0 in 100 cm$_3$ of deionised water. Calcium nitrate solution (0.2 mol dm^{-3}) – dissolve 4.7 g of Ca(NO$_3$)$_2$.4H$_2$O in 100 cm³ of deionised water.

Chromium

Potassium chromate solution (0.2 mol dm^{-3}) – dissolve 3.9 g of K$_2$CrO$_4$ in 100 cm³ of deionised water.

THE ROYAL
SOCIETY OF
CHEMISTRY

Cobalt
Cobalt nitrate solution (0.5 mol dm^{-3}) – dissolve 14.6 g of Co(NO$_3$)$_2$.6H$_2$O in 100 cm^3 of deionised water.

Copper
Copper sulphate solution (0.5 mol dm^{-3}) – dissolve 12.5 g of CuSO$_4$.5H$_2$O in 100cm^3 of deionised water. Copper sulphate solution (0.2 mol dm^{-3}) – dissolve 5.0 g of CuSO$_4$.5H$_2$O in 100cm^3 of deionised water.

Iron
Iron(III) nitrate solution (0.2 mol dm^{-3}) – dissolve 8.1 g of Fe(NO$_3$)$_3$.9H$_2$O in 100 cm^3 of deionised water. Iron(II) sulphate solution (0.2 mol dm^{-3}) – dissolve 5.6 g of FeSO$_4$.7H$_2$O in 100 cm^3 of deionised water. Add sulphuric acid (1 mol dm-3) to make up to 200 cm^3. (The presence of the acid minimises the hydrolysis of iron(II).)

Lead
Lead nitrate solution (0.5 mol dm^{-3}) – dissolve 16.6 g of Pb(NO$_3$)$_2$ in 100 cm^3 of deionised water.

Lithium
Lithium bromide solution (1 mol dm^{-3}) – dissolve 4.3 g of lithium bromide in 50 cm^3 of deionised water.

Magnesium
Magnesium nitrate solution (0.5 mol dm^{-3}) – dissolve 7.4 g of Mg(NO$_3$)$_2$ in 100 cm^3 of deionised water.

Manganese
Potassium manganate(VII) solution (0.01 mol dm^{-3}) – dissolve 0.16 g of KMnO$_4$ in 100 cm^3 of deionised water.

Molybdenum
Ammonium molybdate solution (0.05 mol dm^{-3}) – dissolve 6.2 g of (NH$_4$)$_6$Mo$_7$O$_{24}$.4H$_2$O in 100 cm^3 of water.

Nickel
Nickel nitrate solution (0.5 mol dm^{-3}) – dissolve 14.5 g of Ni(NO$_3$)$_2$.6H$_2$O in 100 cm^3 of deionised water.

Potassium
Potassium bromide solution (0.2 mol dm^{-3}) – dissolve 2.4 g KBr in 100 cm^3 of deionised water. Potassium iodide (0.2 mol dm^{-3}) – dissolve 3.3 g KI in 100 cm^3 of deionised water.

Silver
Silver nitrate solution(0.1 mol dm-3) – dissolve 1.7 g of AgNO$_3$ in 100 cm^3 of deionised water. Store in a dark place.

Sodium
Sodium fluoride solution (0.5 mol dm^{-3}) – dissolve 1.1g of NaF in 50 cm^3 of deionised water. Sodium chloride solution (0.5 mol dm^{-3}) – dissolve 2.9 g of NaCl in

THE ROYAL
SOCIETY OF
CHEMISTRY

100 cm³ of deionised water. Sodium carbonate solution (0.5 mol dm⁻³) – dissolve 5.3 g of Na_2CO_3 in 100 cm³ of deionised water. Sodium sulphate (0.5 mol dm⁻³) – dissolve 7.1 g of Na_2SO_4 in 100 cm³ of deionised water.

Strontium
Strontium nitrate solution (0.5 mol dm⁻³) – dissolve 10.6 g of $Sr(NO_3)_2$ in 100 cm³ of deionised water.

Tungsten
Sodium tungstate solution (0.2 mol dm⁻³) – dissolve 6.6 g of $Na_2WO_4.2H_2O$ in 100 cm³ of deionised water.

Vanadium
Ammonium vanadate solution (0.2 mol dm⁻³) – dissolve 2.3 g NH_4VO_3 in 100 cm³ of deionised water.

Zinc
Zinc sulphate solution (0.2 mol dm⁻³) – dissolve 5.8 g of $ZnSO_4.7H_2O$ in 100 cm³ of deionised water.

Techniques

1. Microscale filtration
This is a simple but effective method for constructing a filter funnel from a plastic pipette.

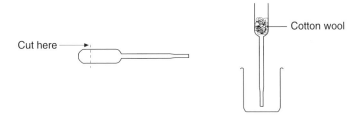

Microscale filtration

Your filter funnel is now ready to use. The efficiency of the filter funnel depends mainly on how compact the cotton wool is in the funnel. For coarse particles the cotton wool need not be packed very tightly. However, if very fine particles are to be separated from the liquid tight packing is essential for effective separation.

NB In microscale filtration, transfer of liquids is always by pipette never by pouring.

2. Sampling a bottle of hydroxybenzene (phenol)
Hydroxybenzene is a hazardous substance and sampling a bottle of hydroxybenzene using a spatula is usually difficult. The need to ensure that crystals of hydroxybenzene do not come into contact with the skin, the wearing of gloves and the fact that hydroxybenzene is hygroscopic, causing the crystals to stick together, all add to the difficulties. The following procedure reduces safety hazards and allows students to gain in confidence and practical skills. Students must still wear eye protection and gloves.

THE ROYAL
SOCIETY OF
CHEMISTRY

This technique illustrates the use of two plastic pipettes for obtaining small samples of hydroxybenzene crystals suitable for use in microscale chemistry experiments.

Procedure

1.   Take a standard form plastic pipette and cut off the ends as indicated.

2.   Cut the tip off the end of a fine-tip pipette as indicated.

3.   Taking the modified standard pipette, press it gently down into the crystals in the hydroxybenzene bottle and withdraw. A small column of solid hydroxybenzene should be held on the inside of the bevelled end.

4.   Place the pipette over the petri dish and insert the fine-tipped pipette to press out a small quantity of hydroxybenzene crystals. (Repeat if necessary at other locations in the petri dish for example if you are doing the 'Reactions of Hydroxybenzene' experiment).

5.   Place the ends of both pipettes into a 100 cm^3 beaker about half full with 1 mol dm^{-3} sodium hydroxide solution. This dissolves any solid hydroxybenzene remaining on the pipettes.

6.   The pipettes may then be washed with deionised water, dried on tissue paper and stored ready for re-use.

Sampling a bottle of hydroxybenzene

THE ROYAL
SOCIETY OF
CHEMISTRY

Microscale titrations

Teacher's guide

Titrations are widely used in post-16 chemistry courses. The traditional apparatus comprises 50 cm³ burettes and 25 cm³ pipettes. For a whole class, experiments with this size of apparatus consume large quantities of solutions and students often take a long time to do the titrations. They are frequently very messy with spillages from pouring solutions and leaking taps on burettes!

Microscale titrations have several advantages compared with conventional titrations. Details of how to construct the apparatus are given in the student worksheet.

Description

In these microscale titrations the 50 cm³ burette is replaced by a 2 cm³ graduated pipette and the 25 cm³ pipette by a 1 cm³ pipette. A diagram of the apparatus is shown in the student worksheet overleaf.

To do the titration the plunger is gently pressed down. The volume of the drops produced is *ca* 0.02 cm³ – without the fine tip the drop volume is *ca* 0.04 cm³. The apparatus can be read to 0.01 cm³ (compared with 0.05 cm³ with a conventional burette) although the volume of solution delivered (less than 2 cm³) is far less.

The main advantages of this technique are:

▼ the greatly reduced volumes of solutions required (with the associated reductions in cost);

▼ the removal of the need for pouring solutions;

▼ increased speed of titration; and

▼ the smaller quantities of solutions to dispose of at the end of the experiment.

The apparatus can be used for the following experiments:

▼ Acid–base neutralisation (p. 63)

▼ Measuring an equilibrium constant (p. 64)

▼ Finding out how much salt there is in seawater (p. 65)

▼ Measuring the amount of vitamin C in fruit drinks (p. 67)

Apparatus (per pair or group)

▼ One clamp stand with two bosses and clamps

▼ One 2 cm³ graduated pipette

▼ One plastic pipette (fine-tip form)

▼ One 10 cm³ plastic syringe

▼ Rubber tubing (for connectors)

▼ Adaptor (made from a 1 cm³ plastic syringe)

▼ One 10 cm³ beaker.

Acknowledgement

I thank Dr Mono M. Singh, Director, National Microscale Chemistry Centre, North Andover, Massachusetts, US for help in preparing this technique.

THE ROYAL
SOCIETY OF
CHEMISTRY

Microscale titrations

Student worksheet

In this technique you will be constructing a microscale titration apparatus and then using it to do titrations.

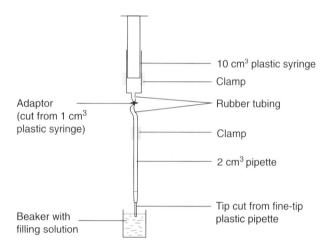

Instructions

1. Place the 2 cm³ pipette in a clamp. This is your 'burette' during the titration.

2. Cut the end off a fine-tip plastic pipette and push it carefully onto the end of the 'burette'.

3. Attach the 10 cm³ plastic syringe to a clamp above the 'burette'.

4. Cut the end off a 1 cm³ plastic syringe to make an adaptor.

5. Cut two short pieces of rubber tubing and use them to attach the syringe to the top of the 'burette' via the adaptor.

THE ROYAL
SOCIETY OF
CHEMISTRY

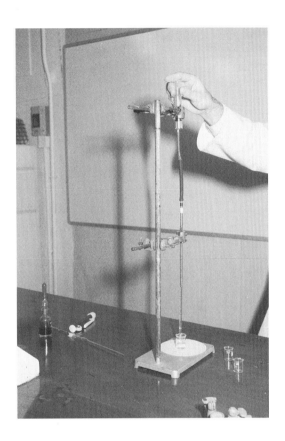

Filling a microscale titration apparatus.

Filling the 'burette'

1. Take a 10 cm³ beaker and using a plastic pipette half-fill it with the solution for the 'burette'.

2. Place the tip of the 'burette' in the solution and slowly raise the plunger. The solution is drawn up into the 'burette'. (If air bubbles are drawn up raise and lower the plunger slowly a few times to expel them). When the desired level is reached release the plunger and the liquid level in the 'burette' should remain stationary. If the level falls, adjust the connections on the apparatus to ensure that the system is air-tight and repeat the filling process.

THE ROYAL
SOCIETY OF
CHEMISTRY

Constructing a conductivity meter

Teacher's guide

This procedure gives instructions for constructing a conductivity meter that can be used for testing conductivity of solutions or solids. It is possible for students to make the apparatus themselves. Alternatively students could be given the instruments ready-made.

Apparatus (per group)

▼   One light-emitting diode (LED)

▼   One 1000 Ω resistor

▼   One 9 V battery

▼   One battery connector

▼   One piece of thin wood or plastic – *ca* 150 x 25 mm

▼   Solder wire or tape

▼   Soldering iron

▼   Sticking tape

▼   One thick rubber band.

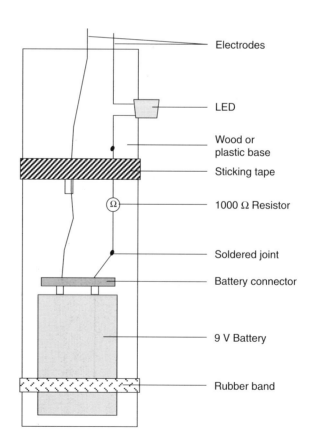

THE ROYAL
SOCIETY OF
CHEMISTRY

Constructing a conductivity meter

Student worksheet

In this experiment you will be making a conductivity meter which you can use to test the conductivity of solutions and solids such as metals.

Instructions

A diagram of the meter is shown here. Your teacher will give you guidance on how to build the conductivity meter.

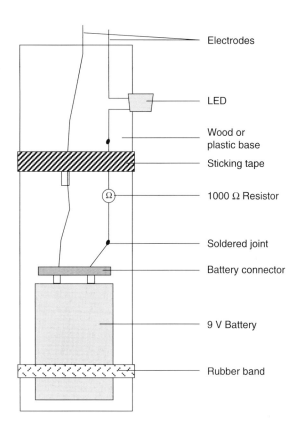

THE ROYAL
SOCIETY OF
CHEMISTRY

Microscale Hoffman apparatus

Teacher's guide

Topic

Electrolysis.

Level

Pre-16 and post-16.

Timing

20 min.

Description

In this experiment students make a microscale Hoffman apparatus from plastic pipettes and use it to investigate aspects of electrolysis. Instructions are given here for the construction of the apparatus. Experiment 38 describes an electrolysis using the apparatus.

Apparatus

▼   Four plastic pipettes (standard form, *eg* Aldrich ref Z13,500-3)

▼   Scissors

▼   A pin

▼   Pencil leads (HB 0.9 mm)

▼   Adhesive, *eg* polystyrene glue.

Construction

1.   Using the scissors cut two holes in two of the pipettes as indicated in the diagram.

2.   Cut off the stem from a third pipette and insert it in the first two pipettes to join them.

3.   Cut off the tip and end of the bulb of the fourth pipette and insert the tip end into a hole in the middle of the stem joining the first two pipettes.

4.   With the pin make a hole in the bottom of the two bulbs and carefully insert a pencil lead into each.

5.   Apply adhesive to each joint and leave to dry.

THE ROYAL
SOCIETY OF
CHEMISTRY

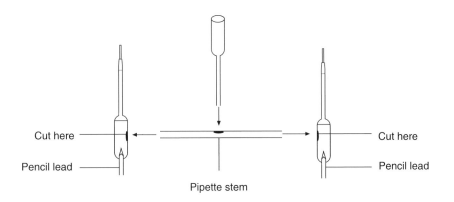

Cut here

Pencil lead

Pipette stem

Cut here

Pencil lead

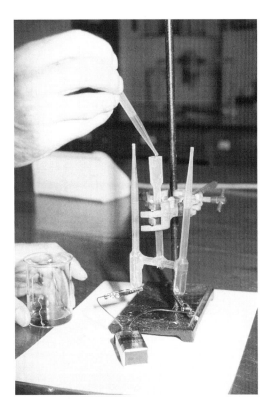

A set-up showing a microscale Hoffman apparatus.

Reference

Chemunity News, 1994, American Chemical Society.

THE ROYAL
SOCIETY OF
CHEMISTRY

Microscale Hoffman apparatus

Student worksheet

Follow these instructions to make a microscale Hoffman electrolysis apparatus out of plastic pipettes.

Construction

1. Using the scissors cut two holes in two of the pipettes as indicated in the diagram.

2. Cut off the stem from a third pipette and insert it in the first two pipettes to join them.

3. Cut off the tip and end of the bulb of the fourth pipette and insert the tip end into a hole in the middle of the stem joining the first two pipettes.

4. With the pin make a hole in the bottom of the two bulbs and carefully insert a pencil lead into each.

5. Apply adhesive to each joint and leave to dry.

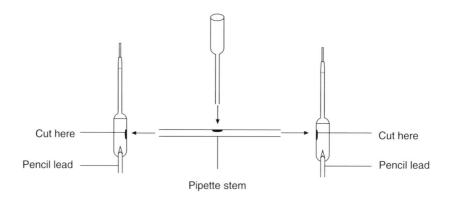

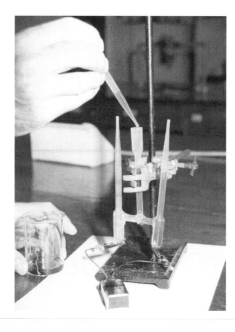

STUDENT WORKSHEETS

THE ROYAL
SOCIETY OF
CHEMISTRY

1. Acids and bases

In this experiment you will be testing various substances with indicator solution and looking for colour changes.

Instructions

1.   Place a clear plastic sheet over the worksheet.

2.   Put two drops of each solution in the appropriate box on the plastic sheet.

3.   Add one drop of full-range indicator to each solution.

Questions

1.   What conclusions can you draw from your observations?

THE ROYAL
SOCIETY OF
CHEMISTRY

Hydrochloric acid	
Sodium hydroxide	
Vinegar	
Sodium carbonate	
Ammonia	
Nitric acid	
Bleach	
Lemon juice	
Sulphuric acid	
Soap solution	

THE ROYAL
SOCIETY OF
CHEMISTRY

2. A chemical reaction

In this experiment you will be looking at the reaction between lead nitrate and potassium iodide. You will be exploring some of the conditions that are necessary for this reaction to take place.

Part A

Instructions

1. Cover the worksheet with a clear plastic sheet.

2. Using tweezers, put two crystals of lead nitrate and two crystals of potassium iodide in the box below.

3. Mix with a straw.

Question

1. Do you notice any changes?

Part B

Follow the instructions below:

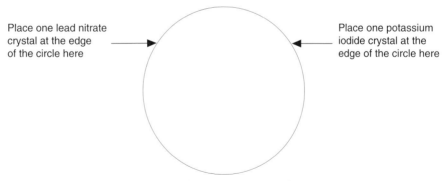

Place one lead nitrate crystal at the edge of the circle here

Place one potassium iodide crystal at the edge of the circle here

Put 10 drops of water in the middle of this circle

Using a straw, carefully push the crystals into the water.

Question

1. What do you observe? Give explanations for your observations.

THE ROYAL
SOCIETY OF
CHEMISTRY

3. Observing chemical changes

In this experiment you will be observing the changes that occur when you mix
solutions of chemicals on the grid shown.

Instructions

1. Cover the worksheet with a clear plastic sheet.

2. Put two drops of barium nitrate solution into box 1 (at the top of the middle
    column). Add two drops of sodium sulphate solution to the drops of barium
    nitrate solution.

3. Put two drops of lead nitrate soluion into box 2. Add two drops of potassium
    iodide solution to the drops of lead nitrate solution.

4. Put two drops of iron(III) nitrate solution into box 3. Add one drop of
    potassium thiocyanate solution to the iron(III) nitrate solution.

5. Put two drops of copper(II) sulphate solution into box 4. Add two drops of
    ammonia solution to the copper(II) sulphate solution.

6. Put two drops of ammonium vanadate(V) solution into box 5. Add one drop of
    hydrochloric acid, then a small piece of zinc metal to the ammonium
    vanadate(V) solution.

7. Put two drops of iron(II) sulphate solution into box 6. Add two drops of sodium
    hydroxide solution to the iron(II) sulphate solution.

8. Put two drops of potassium manganate(VII) solution into box 7. Add two drops
    of iron(II) sulphate solution to the potassium manganate(VII) solution.

9. Put two drops of barium nitrate solution into box 8. Add two drops of sodium
    hydroxide to the barium nitrate solution. Observe, and record any changes
    over the next 10 min.

10. Put one drop of silver nitrate solution into box 9. Add one drop of iron(II)
    sulphate to the silver nitrate solution. Observe closely using a magnifying
    glass.

11. Put two drops of copper(II) sulphate solution into box 10. Add a small piece of
    zinc metal to the copper sulphate solution.

THE ROYAL
SOCIETY OF
CHEMISTRY

Barium nitrate	1	Sodium sulphate
Lead nitrate	2	Potassium iodide
Iron(III) nitrate	3	Potassium thiocyanate
Copper(II) sulphate	4	Ammonia solution
Ammonium vanadate(V)	5	Hydrochloric acid and zinc
Iron(II) sulphate	6	Sodium hydroxide
Potassium manganate(VII)	7	Iron(II) sulphate
Barium nitrate	8	Sodium hydroxide
Silver nitrate	9	Iron(II) sulphate
Copper(II) sulphate	10	Zinc

THE ROYAL
SOCIETY OF
CHEMISTRY

4. The reaction of metals with acids

In this experiment you will be looking at the reactions between various metals and some acids.

Read the instructions before you start the experiment to make sure you understand the procedure.

Instructions

1. Cover the worksheet with a clear plastic sheet.

2. Place a few copper turnings in each box in the copper row.

3. Place one small piece of magnesium ribbon in each box in the magnesium row.

4. Place a few zinc granules in each box in the zinc row.

5. Place some iron filings in each box in the iron row.

6. Finally, place a few tin granules in each box in the tin row.

    When all the pieces of metal are in place:

7. Add two drops of dilute hydrochloric acid to each metal in the hydrochloric acid column.

8. Add two drops of dilute nitric acid to each metal in the nitric acid column.

9. Add two drops of dilute sulphuric acid to each metal in the sulphuric acid column.

10. Finally, put one piece of copper turning in the box at the bottom and add two drops of concentrated nitric acid.

Comments

As you do these experiments observe carefully and record your findings.

Question

1. What do you observe? Give explanations for your observations.

THE ROYAL
SOCIETY OF
CHEMISTRY

	Hydrochloric acid	Nitric acid	Sulphuric acid
Copper			
Magnesium			
Zinc			
Iron			
Tin			

	Concentrated nitric acid
Copper	

THE ROYAL
SOCIETY OF
CHEMISTRY

5. Displacement reactions of metals

In this experiment you will be looking at the reactions between various metals and metal salt solutions.

Instructions

1. Cover the worksheet with a clear plastic sheet.

2. Place a copper turning in each box in the copper row.

3. Place one small piece of magnesium ribbon in each box in the magnesium row.

4. Place a few zinc granules in each box in the zinc row.

5. Place an iron nail in each box in the iron row.

    When all the pieces of metal are in place:

6. Add two drops of copper sulphate(II) solution to each metal in the first column. Observe and record your observations.

7. Add two drops of magnesium nitrate solution to each metal in the second column. Observe and record your observations.

8. Add two drops of zinc chloride solution to each metal in the third column. Observe and record your observations.

9. Finally, add two drops of iron(III) nitrate solution to each metal in the fourth column. Observe and record your observations.

THE ROYAL
SOCIETY OF
CHEMISTRY

	Copper(II) sulphate solution	Magnesium nitrate solution	Zinc chloride solution	Iron(III) nitrate solution
Copper				
Magnesium				
Zinc				
Iron				

Try to place the metals in order of reactivity and write equations for any reactions that you observe.

Questions

1. What is the order of reactivity of the metals?

2. Write equations for any reactions that you observe.

THE ROYAL
SOCIETY OF
CHEMISTRY

6.   Redox reactions

In this experiment you will be observing and interpreting two redox reactions.

A. Investigating the reaction between copper(II) ions and halide ions.

Instructions

1.   Cover the worksheet with a clear plastic sheet.

2.   Put one drop of copper(II) sulphate solution in each of the boxes below.

3.   Add one drop of sodium chloride solution to the first box; one drop of potassium bromide solution to the second box; one drop of potassium iodide solution to the third box. Observe.

4.   Add one drop of starch solution to each of the reaction mixtures. Observe.

	Sodium chloride solution	Potassium bromide solution	Potassium iodide solution
Copper(II) sulphate solution			

Question

1.   What explanation can you give for your observations?

THE ROYAL
SOCIETY OF
CHEMISTRY

B. Investigating the reaction between silver(I) ions and iron(II) ions.

Instructions

1. Cover the worksheet with a clear plastic sheet.

2. Put one drop of silver nitrate solution in the box below.

3. Add one drop of iron(II) solution. Observe closely. What happens?

4. After one minute add one drop of thiocyanate solution.

	1. Iron(II) solution
Silver nitrate solution	
	2. Thiocyanate solution

5. To help you interpret your observations, put one drop of potassium thiocyanate solution in each of the boxes below. Add one drop of each of the reagents indicated and observe.

	Silver nitrate solution	Iron(II) solution	Iron(III) solution
Potassium thiocyanate solution			

Question

1. What explanations can you give for your observations?

THE ROYAL
SOCIETY OF
CHEMISTRY

7. The Periodic Table – solubility of sulphates and carbonates of Groups 1 and 2

In this experiment you will be looking to see whether precipitates form when you add drops of solutions of sulphates or carbonates to drops of solutions of Group 1 or 2 metal ions.

Instructions

1.   Cover the worksheets with a clear plastic sheet.

2.   Put two drops of each of the metal ion solutions in each box of the appropriate row.

3.   Add two drops of each of the anion solutions to the appropriate columns.

4.   Observe and interpret your observations.

Group 1	Solution of sulphate ions	Solution of carbonate ions
Solution of lithium ions		
Solution of sodium ions		
Solution of potassium ions		

THE ROYAL
SOCIETY OF
CHEMISTRY

Group 2	Solution of sulphate ions	Solution of carbonate ions
Solution of magnesium ions		
Solution of calcium ions		
Solution of strontium ions		
Solution of barium ions		

Question

1.   How do you account for your observations?

THE ROYAL
SOCIETY OF
CHEMISTRY

8. The Periodic Table –
properties of Group 2 elements

In this experiment you will be observing and interpreting the changes when drops of
solutions of various anions are added to drops of solutions of Group 2 element
cations.

Instructions

1.   Cover the worksheet with a clear plastic sheet.

2.   Put one drop of magnesium solution into each box in the magnesium ions row.

3.   Repeat using calcium solution in the next row, then strontium solution in the
        next row and barium solution in the last row.

4.   Add one drop of fluoride solution to each drop in the fluoride ions column.
        Observe what happens.

5.   Repeat step 4 using each of the other solutions of anions in the subsequent
        columns.

Observe each reaction carefully and record your observations.

	Fluoride ions	Chloride ions	Bromide ions	Iodide ions	Hydroxide ions	Sulphate ions	Carbonate ions
Magnesium ions							
Calcium ions							
Strontium ions							
Barium ions							

Question

1.   What explanations can you give for your observations?

THE ROYAL
SOCIETY OF
CHEMISTRY

9. The Periodic Table – changes down the Group 7 elements

Each of the halogens forms a monovalent (singly-charged) anion. In this experiment you will be looking at the similarities and differences in some of the properties of these halide ions.

Instructions

1. Cover the worksheet with a clear plastic sheet.

2. Put one drop of each of the halide ion solutions in the appropriate boxes.

3. Add one drop of silver nitrate solution to each of the boxes in the first column.

4. Repeat using lithium bromide solution for the second column and calcium nitrate solution for the third column.

Record all your observations.

Silver nitrate solution	Lithium bromide solution	Calcium nitrate solution
Fluoride ions	Fluoride ions	Fluoride ions
Chloride ions	Chloride ions	Chloride ions
Bromide ions	Bromide ions	Bromide ions
Iodide ions	Iodide ions	Iodide ions

Question

1. What explanations can you give for your observations?

THE ROYAL
SOCIETY OF
CHEMISTRY

10. Lead compounds – precipitation reactions and pigments

Many lead compounds are insoluble and some of them are brightly coloured. In this experiment you will be observing some precipitation reactions of lead ions.

Follow the instructions in both parts of the experiment and record your observations and try to give explanations.

Part A

Instructions

1. Cover the worksheet with a clear plastic sheet.

2. Put one drop of lead nitrate solution in each box.

3. Add one drop of each of the solutions containing the anions indicated to the appropriate box.

	Hydroxide ions	Chloride ions	Bromide ions	Iodide ions	Carbonate ions	Sulphate ions	Chromate ions
Lead ions							

Questions

1. Which of the lead compounds observed appear to be good pigments?

2. What is the main disadvantage of using these compounds as pigments?

Part B

Instructions

1. Put one drop of lead nitrate solution into each box.

2. Add one drop of deionised water and one drop of tap water to the appropriate boxes.

	Deionised water	Tap water
Lead ions		

Question

1. What explanations can you give for your observations?

THE ROYAL
SOCIETY OF
CHEMISTRY

11. The chemistry of silver

In this experiment you will be looking at the properties of silver compounds.

Instructions

1.   Cover the worksheets with a clear plastic sheet.

2.   Put one drop of silver nitrate solution in each box in the top two silver nitrate rows only. (The third row is left empty for now.)

3.   Add one drop of each of the chloride, bromide and iodide solutions to the drop of silver nitrate solution in the appropriate box.

4.   Make a cover from your piece of card and place it over all the drops in the first row only.

5.   Record your observations of the uncovered row.

6.   After 15 min remove the cover and compare the covered and uncovered rows.

7.   Put one drop of silver nitrate solution in each box in the third silver nitrate row.

8.   Add one drop of each of the chloride, bromide and iodide solutions to the drops of silver nitrate solution in the appropriate box.

9.   Add five drops of ammonia solution to each of the drops.

10. Stir gently with the tip of a pipette.

11. Record your observations.

	Chloride ions	Bromide ions	Iodide ions
Silver nitrate solution			
Silver nitrate solution			
Silver nitrate solution			

12. Put one drop of silver nitrate solution in the box below.

13. Add one drop of iron(II) solution.

14. Observe closely using a magnifying glass.

	Iron(II) solution
Silver nitrate solution	

THE ROYAL
SOCIETY OF
CHEMISTRY

12. Iron chemistry –
variable oxidation state

The purpose of this experiment is to compare the chemistry of the two main oxidation states of iron (a first row transition element) and to consider explanations for any differences observed. Carefully follow the instructions below noting down all your observations and trying to give explanations.

Instructions

1. Cover the worksheet with a clear plastic sheet.

2. Put one drop of iron(II) solution in each box in the second row.

3. Put one drop of iron(III) solution in each box in the third row.

4. Add two drops of sodium hydroxide solution to each drop in the boxes in the second column. Observe and note whether there are any changes over the next 10 min.

5. Add one drop of potassium thiocyanate solution to each drop in the third column.

6. Add one drop of potassium iodide solution to each drop in the fourth column. After one minute, add one drop of starch solution to each.

7. Add one drop of potassium manganate(VII) solution to each drop in the fifth column. Observe changes over the next 10 min.

8. Add one drop of silver nitrate solution to each drop in the sixth column. Observe closely using a magnifying glass.

Solutions of	Hydroxide ions	Thiocyanate ions	Iodide ions	Manganate (VII) ions	Silver(I) ions
Iron(II) ions					
Iron(III) ions					

Questions

1. What explanations can you give for your observations?

2. Can you write equations for the reactions you observe?

THE ROYAL
SOCIETY OF
CHEMISTRY

13. The transition elements

The purpose of this experiment is to examine some of the solution chemistry of the transition elements.

In particular you will be looking for evidence of complex formation and change in oxidation state – two important general characteristics of transition elements.

Instructions

1. Cover the worksheet with a clear plastic sheet.

2. Put two separate drops of the solutions of each of the elements from vanadium to zinc in the appropriate boxes. Observe and comment.

    Do the experiments for each solution of each element as described below on one of the drops in each box only (the other drop will act as a reference). In each case, observe carefully and try to give explanations for your observations:

Vanadium (V)	–	add one drop of dilute hydrochloric acid and a small piece of zinc.
Chromium (Cr)	–	add one drop of silver nitrate solution.
Manganese (Mn)	–	add one drop of iron(II) solution.
Iron (Fe)	–	add one drop of potassium iodide solution. After one minute add one drop of starch test solution.
Cobalt (Co)	–	add one drop of ammonia solution.
Nickel (Ni)	–	add two drops of sodium hydroxide solution.
Copper (Cu)	–	add one drop of ammonia solution.
Zinc (Zn)	–	add two drops of sodium hydroxide solution.

Solutions of transition metal ions	V	Cr	Mn	Fe	Co	Ni	Cu	Zn

Questions

1. Which element among the ones that you have tested does not behave as a transition element?

2. Why is this?

THE ROYAL
SOCIETY OF
CHEMISTRY

14. Reactions of transition elements

The purpose of this experiment is to observe and interpret some of the chemistry of three first row transition elements and to compare them with a typical s-block element. Follow the instructions carefully recording all your observations.

Instructions

1. Cover the worksheet with a clear plastic sheet.

2. Place two drops of each of the solutions of cobalt, nickel, copper and magnesium ions in the appropriate boxes in the column headed Metal ion solution.

	Metal ion solution	Ammonia solution	Sodium hydroxide solution
Solution of cobalt ions			
Solution of nickel ions			
Solution of copper ions			
Solution of magnesium ions			

Reactions with ammonia

1. In the column headed ammonia solution put four drops of each element solution in two separate lots either side of the dashed line.

2. For cobalt, add one drop of ammonia solution. Add three further drops of ammonia solution to the right hand side of the dashed line only.

3. For nickel, add one drop of ammonia solution. After one minute add five further drops of ammonia to the right hand side of the dashed line.

4. For copper, add one drop of ammonia solution and stir with the tip of a pipette. Add three further drops of ammonia solution to the right hand side of the dashed line.

5. For magnesium add two drops of ammonia solution. Add two further drops of ammonia solution to the right hand side of the dashed line.

Reactions with sodium hydroxide

1. Put two drops of each metal ion solution in the boxes in the column headed sodium hydroxide solution.

2. Add two drops of sodium hydroxide solution to each.

Question

1. How do the reactions of the transition metal ions differ from those of the s-block metal ions?

15. Chromium, molybdenum and tungsten

In this experiment you will be looking at a group of transition elements chromium, molybdenum and tungsten. Follow the instructions below recording all your observations and try to give explanations where possible.

Instructions

1. Cover the worksheet with a clear plastic sheet.

2. Put one drop of each of the metal ion solutions in the appropriate boxes in the column headed ion. Observe and comment.

3. Put one drop of each of the metal ion solutions in the appropriate boxes in the remaining two columns.

4. Add one drop of hydrochloric acid to each drop in the column headed hydrochloric acid solution. Observe carefully, noting any changes over the next few minutes.

5. Add one drop of iron(II) solution to each drop in the column headed solution of iron(II) ions.

6. Add one drop of sodium hydroxide solution to the chromium drop in the box in the hydrochloric acid solution column. Observe any changes.

Element	Ion	Hydrochloric acid solution	Solution of iron(II) ions
Chromium (Cr)	CrO_4^{2-}		
Molybdenum (Mo)	$Mo_7O_{24}^{6-}$		
Tungsten (W)	WO_4^{2-}		

Questions

1. Using a data book if necessary, can you explain why these three elements are placed in the same group in the transition metal block? (Hint: Consider how the atomic structures of the alkali metals are similar to each other.)

2. Can you write an equation to explain the colour changes observed on adding acid or alkali to the chromium solution?

THE ROYAL
SOCIETY OF
CHEMISTRY

16. The reaction between hydrogen peroxide and dichromate ions

In this experiment you will be looking at the reaction between two substances containing oxygen.

Instructions

1. Cover the worksheet with a clear plastic sheet.

2. Put one drop of potassium dichromate solution in the circle below.

3. Add one drop of 5% hydrogen peroxide to the potassium dichromate solution.

Questions

1. Observe carefully. Are there any changes over the next few minutes?

2. Give an explanation for your observations and try to write an equation for the reaction. What type of reaction is occurring?

THE ROYAL
SOCIETY OF
CHEMISTRY

17. The chemistry of thiosulphate ions

In this experiment you will be looking at some interesting chemical reactions of sodium thiosulphate. You will probably already be aware of the reaction between sodium thiosulphate and iodine.

Part A　　**The reaction between thiosulphate ions and iodine solution**

Instructions

1.　　Cover the worksheets with a clear plastic sheet.

2.　　Put one drop of iodine solution in the box below.

3.　　Add two drops of thiosulphate solution.

	Solution of aqueous iodine
Solution of thiosulphate ions	

Observe, comment and write an equation for the reaction.

Question

1.　　What type of reaction are you observing?

Part B　　**The reaction between thiosulphate and silver halide**

Instructions

1.　　To form the silver halides, first put one drop of silver nitrate solution into each of the empty boxes below, then add one drop of potassium bromide solution and potassium iodide solutions into the appropriate boxes. Observe and comment.

2.　　Add three drops of sodium thiosulphate solution to each box and stir with the end of a pipette. Observe and comment.

	1. Potassium bromide solution	1. Potassium iodide solution
Silver nitrate solution		
	2. Sodium thiosulphate solution	2. Sodium thiosulphate solution

THE ROYAL
SOCIETY OF
CHEMISTRY

Question

1.   What explanations can you give for your observations?

Part C   The reaction between
thiosulphate ions and transition metal ions

Instructions

1.   Put two drops of iron(III) solution in the first box.

2.   Put two drops of iron(III) solution and one drop of copper(II) solution in the second box.

3.   Put two drops of copper(II) solution in the third box.

4.   Add one drop of thiosulphate solution to each box and observe carefully, especially the second box.

Solution of	Iron(III)ions	Iron(III) + Copper(II) ions	Copper(II) ions
Thiosulphate ions			

Question

1.   What explanations can you give for your observations?

THE ROYAL
SOCIETY OF
CHEMISTRY

18. The equilibrium of the cobalt chloride–water system

In this experiment you will be looking at the reversible reaction shown in the equation below. The hexaquocobalt(II) complex is pink while the tetrachlorocobalt(II) complex is blue. The different colours of the reactants and products allow you to observe the effect on the position of equilibrium of the addition of water and chloride ions and the removal of water.

$$Co(H_2O)_6^{2+} + 4Cl^- \rightleftharpoons CoCl_4^{2-} + 6H_2O$$
Pink   Blue

Instructions

Follow the instructions carefully recording the colours of the solutions at each stage. Appropriate care should be taken when using concentrated acids.

1. Put 50 drops of ethanolic cobalt chloride solution into each of the wells A1–A6 and B1 (see diagram below). Place 20 drops of aqueous cobalt chloride into wells B2–B4.

2. Add eight drops of water to wells A2–A5 and carefully swirl the well-plate for 1 min.

3. Add eight drops of potassium chloride solution to wells A6 and B1. Swirl the well-plate gently for 1 min.

4. Add 10 drops of concentrated hydrochloric acid to well A3 and enough solid potassium chloride to wells A4 and A5 to cover the bottoms of the wells. Swirl carefully for 1 min.

5. Add 20 drops, five drops at a time with gentle swirling, of concentrated sulphuric acid to wells A5–B1.

6. Add 20 drops of concentrated hydrochloric acid to wells B3 and B4, swirling gently.

7. Finally, add 30 drops of water to well B4.

Questions

1. Can you give explanations for all your observations – writing equations where appropriate?

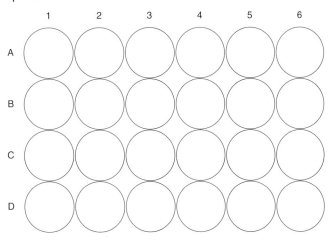

THE ROYAL
SOCIETY OF
CHEMISTRY

19. Mass changes in chemical reactions

In this experiment you will be doing two chemical reactions to see whether any mass changes occur.

Instructions

Part A The reaction between sodium carbonate and calcium nitrate

1. Put two plastic pipettes containing the solutions of sodium carbonate and calcium nitrate in the outer two wells of the mini well-plate (see below).

2. Place on a balance and record the mass.

3. Put 20 drops of sodium carbonate solution into the middle well followed by 20 drops of calcium nitrate solution.

4. Record any changes you see and write an equation for the reaction.

5. Reweigh the complete apparatus and record the mass.

Question

1. Is there a difference in the masses before and after the reaction? Explain your answer.

Part B The reaction between marble and hydrochloric acid

1. Place one piece of marble chip and the pipette containing the hydrochloric acid in two of the wells in the mini well-plate.

2. Add 10 drops of hydrochloric acid to the well containing the marble chip.

3. Record any changes you see and write an equation for the reaction.

4. When the reaction has finished reweigh the complete apparatus and record the mass.

Question

1. How do your answers compare with those in Part A? Explain your answers.

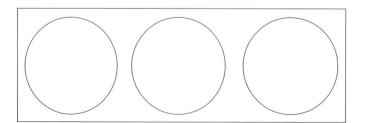

THE ROYAL
SOCIETY OF
CHEMISTRY

20. Measuring density

In this experiment you will be measuring the mass and volume of seawater and tap
water and then using your data to determine the density. (Density = mass / volume)

Instructions

1. Place the measuring cylinder on the balance pan and tare the balance.

2. Carefully add 0.5 cm³ of tap water dropwise to the measuring cylinder. Record
    in a table the volume of water added and the mass.

3. Add drops of tap water until the volume is 1.0 cm³. Record the new mass.

4. Add water until the volume is 1.5 cm³ and record the mass.

5. Continue in this manner at 0.5 cm³ intervals up to 5.0 cm³.

    (If you cannot tare the balance, subtract the mass of the measuring cylinder
    each time.)

Repeat the whole process using seawater.

Questions

Make a table of your results (as shown below).

Seawater				Tap water		
Vol (cm³)	Mass (g)	Density (g cm⁻³)		Vol (cm³)	Mass (g)	Density (g cm⁻³)

On a piece of graph paper plot the volume against the mass for both tap water and
seawater and draw a best line fit through each set of points. Measure the slope of
these lines.

Questions

1. What do you notice on measuring the slope of the lines on your graph?

2. What are the advantages of showing your results graphically rather than just in
    a table? (You may wish to use a spreadsheet package to do the calculations.)

THE ROYAL
SOCIETY OF
CHEMISTRY

21. Energy changes in neutralisation

In this experiment you will be using a special temperature strip to study energy changes in two chemical reactions.

Instructions

1. Put two drops of dilute hydrochloric acid on the strip in each of the places marked by the arrows.

2. Add a small piece of magnesium ribbon to each drop.

3. Observe and explain your findings.

4. Wipe the strip clean with tissue paper.

5. Put one drop of hydrochloric acid on the strip in each of the places marked by the arrows.

6. Add one drop of sodium hydroxide to each drop of acid.

7. Observe and explain your findings.

THE ROYAL
SOCIETY OF
CHEMISTRY

22. Investigating temperature changes on evaporation of liquids

In this experiment you will be using a thermometer strip to examine temperature changes when you put drops of liquids on the strip.

Instructions

1. Put a row of drops of water along the strip. Note the shape of the drops and note whether there are any temperature changes over the next few minutes.

2. Repeat using ethanol.

3. Repeat using ethoxyethane.

4. Record your observations in a table and try to give explanations. Bear in mind what you know about intermolecular forces when you interpret your findings.

Question

1. If you have been swimming and you do not dry yourself quickly after you get out of the water you often start to feel cold. Why do you think this is?

THE ROYAL
SOCIETY OF
CHEMISTRY

23. Investigating the effect of concentration on the rate of a chemical reaction

In this experiment you will be studying the reaction between sodium thiosulphate solution and hydrochloric acid solution which react to produce a fine precipitate of sulphur according to the following equation:

$$S_2O_3^{2-}(aq) + 2H^+(aq) \rightarrow H_2O(l) + SO_2(g) + S(s)$$

You will be varying the concentrations of thiosulphate ions and hydrochloric acid to see what effect these have on the rate at which the precipitate of sulphur is formed. Place the well-plate onto the grid below with the array of crosses under wells A1–A6:

Part A

The effect of varying the concentration of thiosulphate ions

Add drops of the solutions to each of the wells A1–A6 as follows:

Well no	A1	A2	A3	A4	A5	A6
Hydrochloric acid	20	20	20	20	20	20
Water	25	20	15	10	5	0

Now start the stopclock and add 30 drops of thiosulphate solution to well A6 and measure the time until you can no longer see the cross under well A6. Record this time. Repeat this procedure with the other wells adding drops of thiosulphate according to the following table:

Well no	A1	A2	A3	A4	A5	A6
Thiosulphate solution	5	10	15	20	25	30

Question

1.   From your results, plot a graph of the concentration of thiosulphate ions in each well against the time taken for the cross to disappear. Why is your graph a particular shape and what information does this tell you?

THE ROYAL
SOCIETY OF
CHEMISTRY

Part B

Using a similar procedure devise and do an experiment to investigate the effect of varying the concentration of hydrochloric acid on the rate of this reaction.

THE ROYAL
SOCIETY OF
CHEMISTRY

Information sheet for experiments 24 – 30

These experiments all involve generating small amounts of gas inside a petri dish. Then, using test solutions or solid powders placed around the edge of the dish to test for the gas, you observe any colour changes that take place.

Only very small amounts of gas are produced in each experiment (in most cases you will not even see any bubbles of gas!). Provided the instructions are followed carefully, all the gas is contained within the petri dish. These microscale experiments therefore enable you to do chemical reactions that would be difficult and hazardous to do on a larger scale.

Before you start you will need to cut the bottom off a plastic pipette using a pair of scissors. The bottom piece of the pipette will be your reaction vessel in your experiments:

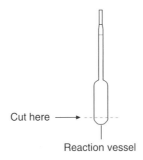

Cut here ⟶

Reaction vessel

At the end of each experiment mop up the chemicals inside the petri dish and the reaction vessel with tissue paper. The dish and vessel can then be re-used.

THE ROYAL
SOCIETY OF
CHEMISTRY

24. Some reactions of ammonia

Instructions

1. Cover the worksheet with a clear plastic sheet.

2. Place the base of the petri dish directly over the circle below. Place the reaction vessel in the centre.

3. At the corners of the triangle add drops of the test solutions only as indicated below (Care: Nessler's reagent is toxic – it contains mercury compounds – make sure that you do not get any on your skin. If you do, wash it off quickly with water).

4. Put three drops of ammonia solution into the reaction vessel and quickly replace the lid on the petri dish.

5. Record all your observations over the next 15 min.

Question

1. What explanations can you give for your observations?

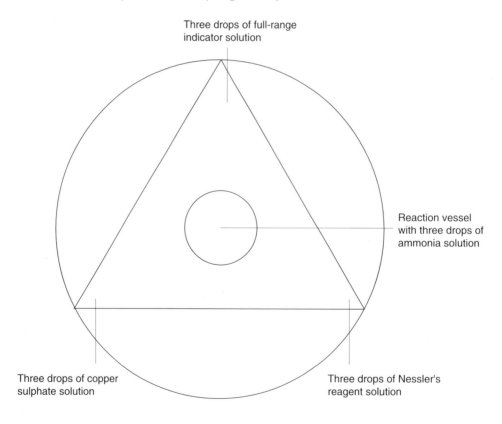

Three drops of full-range indicator solution

Reaction vessel with three drops of ammonia solution

Three drops of copper sulphate solution

Three drops of Nessler's reagent solution

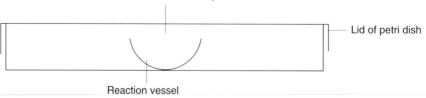

There must be a gap between the top of the reaction vessel and the lid of the petri dish

Lid of petri dish

Reaction vessel

THE ROYAL
SOCIETY OF
CHEMISTRY

25. Some reactions of carbon dioxide

Instructions

1. Cover the worksheet with a clear plastic sheet.

2. Place the base of the petri dish directly over the circle below. Place the reaction vessel in the centre.

3. At the corners of the triangle add drops of the test solutions as indicated below (Care: barium nitrate is toxic).

4. Put a small marble chip in the reaction vessel and add three drops of hydrochloric acid. Quickly replace the lid on the petri dish.

5. Record all your observations over the next 15 min.

Question

1. What explanations can you give for your observations?

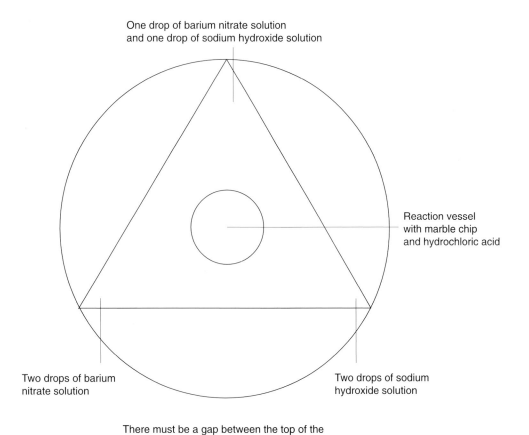

One drop of barium nitrate solution and one drop of sodium hydroxide solution

Reaction vessel with marble chip and hydrochloric acid

Two drops of barium nitrate solution

Two drops of sodium hydroxide solution

There must be a gap between the top of the reaction vessel and the lid of the petri dish

Lid of petri dish

Reaction vessel

THE ROYAL
SOCIETY OF
CHEMISTRY

26. Some reactions of chlorine

Instructions

1. Cover the worksheet with a clear plastic sheet.

2. Place the base of the petri dish directly over the circle below. Place the reaction vessel in the centre.

3. At the corners of the triangle add drops of the test solutions as indicated below.

4. Add two drops of bleach to the reaction vessel followed by three drops of hydrochloric acid. Quickly replace the lid on the petri dish.

5. Record all your observations over the next 15 min. When you have finished, add three drops of sodium hydroxide solution to the reaction vessel to stop the reaction.

Questions

1. What explanations can you give for your observations?

2. Why does sodium hydroxide stop the reaction?

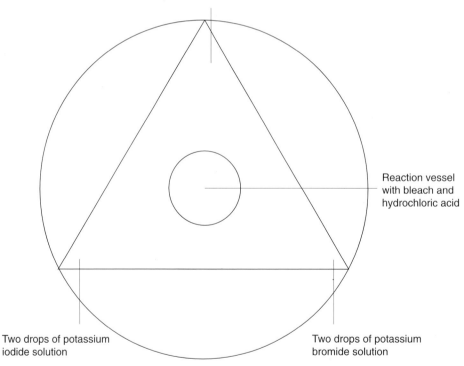

Two drops of sodium chloride solution

Reaction vessel
with bleach and
hydrochloric acid

Two drops of potassium
iodide solution

Two drops of potassium
bromide solution

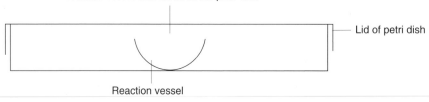

There must be a gap between the top of the
reaction vessel and the lid of the petri dish

Lid of petri dish

Reaction vessel

THE ROYAL
SOCIETY OF
CHEMISTRY

27. More reactions of chlorine

Instructions

1. Cover the worksheet with a clear plastic sheet.

2. Place the base of the petri dish directly over the circle below. Place the reaction vessel in the centre.

3. At the corners of the triangle place small quantities of the solid test powders (using your plastic pipette as a scoop) as indicated below.

4. Add two drops of bleach to the reaction vessel followed by three drops of hydrochloric acid. Quickly replace the lid on the petri dish.

5. Record all your observations over the next 15 min. When you have finished, add three drops of sodium hydroxide solution to the reaction vessel to stop the reaction.

Questions

1. What explanations can you give for your observations?

2. Why does sodium hydroxide stop the reaction?

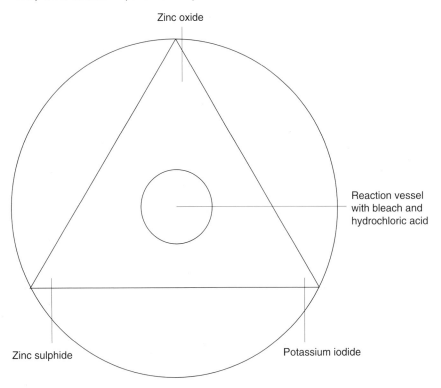

Zinc oxide

Reaction vessel
with bleach and
hydrochloric acid

Zinc sulphide

Potassium iodide

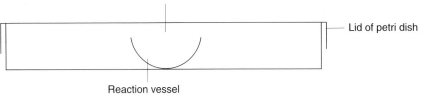

There must be a gap between the top of the
reaction vessel and the lid of the petri dish

Lid of petri dish

Reaction vessel

THE ROYAL
SOCIETY OF
CHEMISTRY

28. Some reactions of hydrogen sulphide

Instructions

1. Cover the worksheet with a clear plastic sheet.

2. Place the base of the petri dish directly over the circle below. Place the reaction vessel in the centre.

3. At the corners of the triangle add drops of the test solutions as indicated below.

4. Add a small quantity of zinc sulphide powder to the reaction vessel followed by three drops of hydrochloric acid. Quickly replace the lid on the petri dish.

5. Record all your observations over the next 15 min. When you have finished add three drops of sodium hydroxide solution to the reaction vessel to stop the reaction.

Questions

1. What explanations can you give for your observations?

2. Why does sodium hydroxide stop the reaction?

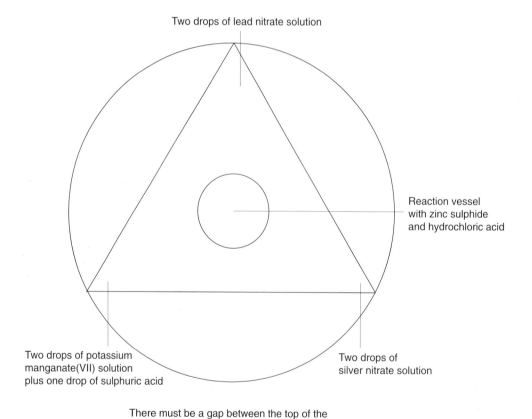

Two drops of lead nitrate solution

Reaction vessel
with zinc sulphide
and hydrochloric acid

Two drops of potassium
manganate(VII) solution
plus one drop of sulphuric acid

Two drops of
silver nitrate solution

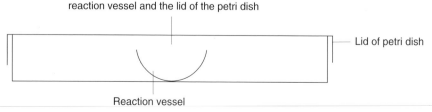

There must be a gap between the top of the
reaction vessel and the lid of the petri dish

Lid of petri dish

Reaction vessel

THE ROYAL
SOCIETY OF
CHEMISTRY

29. Some reactions of nitrogen dioxide

Instructions

1. Cover the worksheet with a clear plastic sheet.

2. Place the base of the petri dish directly over the circle below. Place the reaction vessel in the centre.

3. Put two drops of full-range indicator where shown.

4. At another corner of the triangle place two drops of ammonia solution. Place the lid on the petri dish and wait for the indicator drop to change colour.

5. Remove the lid from the petri dish and, using a piece of tissue, mop up the drop of ammonia.

6. At the two remaining corners of the triangle add the two other test solutions.

7. Add a few copper metal turnings to the reaction vessel followed by three drops of nitric acid. Quickly replace the lid on the petri dish.

8. Record all your observations over the next 15 min.

Question

1. What explanations can you give for your observations.

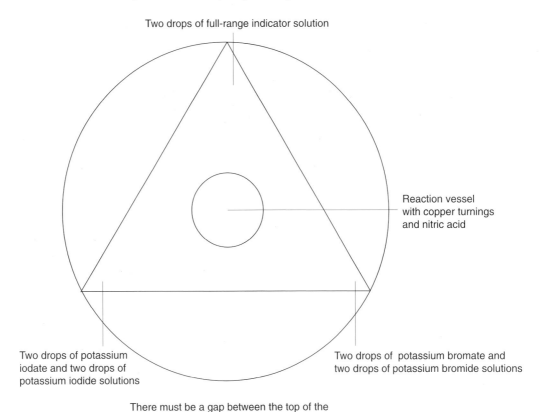

Two drops of full-range indicator solution

Reaction vessel with copper turnings and nitric acid

Two drops of potassium iodate and two drops of potassium iodide solutions

Two drops of potassium bromate and two drops of potassium bromide solutions

There must be a gap between the top of the reaction vessel and the lid of the petri dish

Lid of petri dish

Reaction vessel

THE ROYAL
SOCIETY OF
CHEMISTRY

30. Some reactions of sulphur dioxide

Instructions

1. Cover worksheet with a clear plastic sheet.

2. Place the base of the petri dish directly over the circle below. Place the reaction vessel in the centre.

3. Place two drops of full-range indicator solution where shown.

4. At another corner of the triangle place two drops of ammonia solution. Place the lid on the petri dish and wait for the indicator drop to change colour.

5. Remove the lid from the petri dish and, using a piece of tissue, mop up the drop of ammonia.

6. At the two remaining corners of the triangle add the two other test solutions.

7. Add a small quantity of sodium sulphite powder to the reaction vessel followed by three drops of hydrochloric acid. Quickly replace the lid.

8. Record all your observations over the next 15 min.

Question

1. What explanations can you give for your observations?

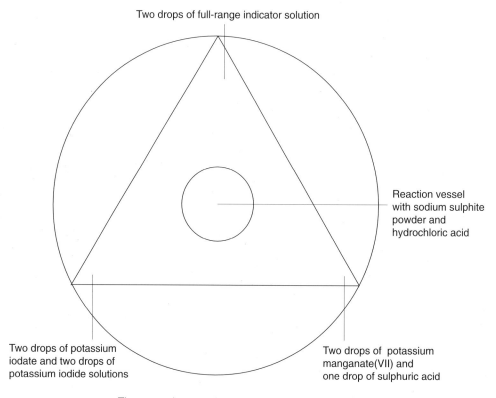

Two drops of full-range indicator solution

Reaction vessel with sodium sulphite powder and hydrochloric acid

Two drops of potassium iodate and two drops of potassium iodide solutions

Two drops of potassium manganate(VII) and one drop of sulphuric acid

There must be a gap between the top of the reaction vessel and the lid of the petri dish

Lid of petri dish

Reaction vessel

THE ROYAL
SOCIETY OF
CHEMISTRY

31. Oxygen and methylene blue

In this experiment you will be generating oxygen gas by reacting hydrogen peroxide and potassium manganate(VII) and testing for it using methylene blue solution.

You will be familiar with testing for oxygen using a glowing splint which re-lights in the gas. This microscale experiment provides an alternative test.

Instructions

1. Construct the gas generating apparatus by cutting the top off a plastic pipette so that a piece of rubber tubing can be attached to the pipette as shown:

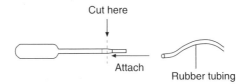

2. Add methylene blue solution to a 10 cm³ beaker until it is about half-full.

3. Add 10 drops of hydrogen peroxide to the shortened pipette.

4. Turn the pipette almost to the horizontal position and carefully put five drops of potassium manganate(VII) solution in the stem as shown:

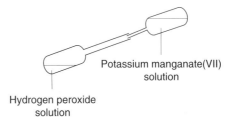

5. Attach the rubber tubing to the pipette, place the other end in the methylene blue solution and gently turn the pipette upright. The potassium manganate(VII) solution, held in the stem, should fall down into the hydrogen peroxide causing vigorous evolution of oxygen gas.

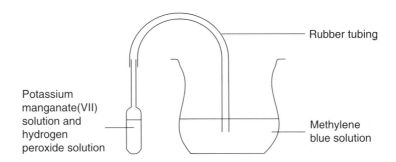

6. Describe your observations.

THE ROYAL
SOCIETY OF
CHEMISTRY

Questions

1. The reaction below shows the structures of methylene blue in the reduced (colourless) and blue (oxidised) forms. Which structure is which? Can you give reasons for your answer?

2. Can you write an equation for the reaction between potassium manganate(VII) and hydrogen peroxide?

THE ROYAL
SOCIETY OF
CHEMISTRY

32. A microscale study of gaseous diffusion

In this experiment you will be observing the diffusion of the gases ammonia and chlorine. You will be doing this by looking for colour changes as the gases react with drops of test solutions.

Instructions

1. Cover the worksheet with a clear plastic sheet.

2. Place two drops of copper(II) sulphate solution in each square (except the one with the circle) of the left hand grid (the one labelled ammonia).

3. Place one drop of potassium iodide solution in each square (except the one with the circle) of the right hand grid (labelled chlorine). Add one drop of starch solution to each drop.

4. Cut the bottom off two plastic pipettes to make a small vessel and place each on the square with the circle.

5. Carefully put four drops of ammonia in the vessel in the 'ammonia' grid and quickly place a well-plate lid over the grid.

6. Carefully put two drops of bleach and two drops of hydrochloric acid in the vessel in the 'chlorine' grid and quickly place a well-plate lid over the grid.

7. Record all your observations over the next 20 min and give explanations.

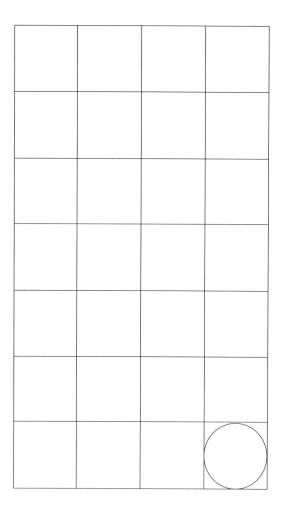

Ammonia

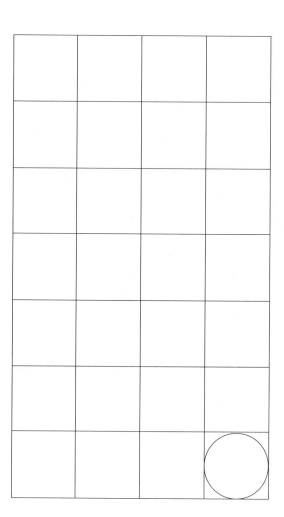

Chlorine

THE ROYAL
SOCIETY OF
CHEMISTRY

33. Acid – base neutralisation

In this experiment you will be using a microscale titration apparatus to do a titration.

Instructions

1. Set up the microscale titration apparatus (see p. 8).

2. Fill the apparatus with dilute hydrochloric acid (see p. 9).

3. Using a 1 cm^3 pipette, add 1 cm^3 of sodium hydroxide solution to a 10 cm^3 beaker.

4. Add 1 drop of phenolphthalein indicator solution.

5. Do the titration by very gently pressing down the plunger on the syringe at the top of the apparatus.

6. Continue until the solution in the beaker is just permanently light pink.

7. Record the volume of hydrochloric acid used in the titration.

8. Repeat the titration until you get reproducible answers.

Calculations

The equation for the neutralisation reaction is:

$$HCl(aq) + NaOH(aq) \rightarrow NaCl(aq) + H_2O(l)$$

From this equation you will see that one mole of hydrochloric acid reacts with one mole of sodium hydroxide (or x moles reacts with x moles).

1. Determine the average value of the volume of hydrochloric acid used in your titrations (let this value be v cm^3).

2. Calculate the number of moles of hydrochloric acid used using the formula:

   $(v/1000)$ x C

   where C is the concentration of the hydrochloric acid (mol dm^{-3})

3. Hence determine the number of moles of sodium hydroxide present.

4. What volume of sodium hydroxide did you use?

5. Hence determine the number of moles of sodium hydroxide present in 1 dm^3.

6. The result that you obtained in step 5 is therefore the concentration of your sodium hydroxide in mol dm^{-3}.

THE ROYAL
SOCIETY OF
CHEMISTRY

34. Measuring an equilibrium constant

In this experiment you will be using your microscale titration apparatus to determine the equilibrium constant for the reaction between silver(I) and iron(II) ions:

$$Ag^+(aq) \ + \ Fe^{2+}(aq) \rightarrow Ag(s) \ + \ Fe^{3+}(aq)$$

Instructions

1. Using a 2 cm³ pipette transfer 2 cm³ each of the 0.10 mol dm⁻³ silver nitrate solution and 0.10 mol dm⁻³ iron(II) sulphate solution to the flask and stopper it so that it is air-tight. Shake the flask and leave undisturbed overnight.

2. Set up the microscale titration apparatus (see p. 8).

3. Fill the apparatus with potassium thiocyanate solution (see p. 9).

4. Using a 1 cm³ pipette, transfer 1 cm³ of the solution to a 10 cm³ beaker, trying not to disturb the silver precipitate.

5. Using the microscale titration apparatus titrate with potassium thiocyanate solution. A permanent red colour marks the end-point.

6. Repeat the titration and calculate the average of your titres.

Calculations

1. The purpose of this experiment is to calculate the equilibrium constant K_c in the expression:

$$K_c \ = \ \frac{[Fe^{3+}(aq)]_{eq}}{[Ag^+(aq)]_{eq} \ [Fe^{2+}(aq)]_{eq}}$$

2. First, work out the initial concentrations of both $Ag^+(aq)$ and $Fe^{2+}(aq)$ in the reaction vessel.

3. Write the equation for the titration reaction and then use it to calculate $[Ag^+]_{eq}$ from your titration results.

4. Since the initial concentrations of both Ag(I) and Fe(II) are equal, it follows that their concentrations at equilibrium are also equal. Therefore you also know $[Fe^{2+}]_{eq}$.

5. Before you can calculate K_c you need to know the concentration of iron(III) at equilibrium, $[Fe^{3+}(aq)]_{eq}$. This is equal to the initial concentration of $Fe^{2+}(aq)$ minus the concentration of $Fe^{2+}(aq)$ at equilibrium. Hence calculate K_c.

THE ROYAL
SOCIETY OF
CHEMISTRY

35. Finding out how much salt there is in seawater

In this experiment you will be using the microscale titration apparatus to find out how much salt there is in seawater. You will be titrating silver nitrate solution against seawater using potassium chromate as indicator.

Instructions

1. Set up the microscale titration apparatus (see p. 8).

2. Fill the apparatus with silver nitrate solution (see p. 9).

3. Using a pipette add 1 cm^3 of seawater to a 10 cm^3 beaker.

4. Add one drop of potassium chromate indicator solution.

5. Titrate until a permanent red colour is observed and record the titre and any other observations.

6. Repeat and take the mean titre.

7. Calculate the salt content of the seawater. (The information below will help you with the calculations.)

Results and calculations

1. What do you think a reasonable answer for the percentage of salt in seawater is likely to be? Compare this with the value you get after working through the following calculations.

2. The equation for the reaction that occurs during the titration is:

   $AgNO_3$ (aq) + $NaCl$(aq) → $AgCl$(s) + $NaNO_3$(aq)

   What is the white precipitate that you see during the titration?

   What causes the red colour at the end-point?

   (Hint: think of some of the characteristics of transition element compounds)

3. From the equation it follows that:

   1 mol of $AgNO_3$ ≡ 1 mol Ag^+ ≡ 1 mol Cl^-

   Calculate the number of moles of silver nitrate used in the titration using the formula:

   $t/1000$ x C

   where t = mean titre volume in cm^3

   and C = concentration of silver nitrate

4. Hence find the number of moles of chloride ion present in 1 cm^3 of seawater.

5. Convert your answer in (3) to a mass by multiplying by the relative atomic mass of chlorine.

6. Convert you answer for (4) to a mass of sodium chloride and hence a percentage of sodium chloride in the seawater. What assumption are you making in your calculation?

THE ROYAL
SOCIETY OF
CHEMISTRY

7.   Compare your result with your earlier guess. Are they similar?

8.   Compare your results with others in your group.

THE ROYAL
SOCIETY OF
CHEMISTRY

36. Measuring the amount of vitamin C in fruit drinks

In this experiment you will be finding out how much vitamin C there is in a fruit drink. The chemical name for vitamin C is ascorbic acid.
The basis of the experiment is as follows.

A known amount of iodine is generated by the reaction between iodate, iodide and sulphuric acid:

$$IO_3^-(aq) + 5I^-(aq) + 6H^+(aq) \rightarrow 3I_2(s) + 3H_2O(l)$$

A measured amount of fruit drink is added. The ascorbic acid in the drink reacts quantitatively with some of the iodine as the iodine is in excess:

HOH$_2$C — C — O — O ... OH ... HO OH Ascorbic acid + I$_2$ ---> HOH$_2$C — C — O — O ... OH ... O O + 2H$^+$ + 2I$^-$ Dehydroascorbic acid

The excess iodine is then titrated against standard sodium thiosulphate solution:

$$I_2 (aq) + 2S_2O_3^{2-}(aq) \rightarrow S_4O_6^{2-}(aq) + 2I^-(aq)$$

From the titration results the amount of iodine that reacts with the sodium thiosulphate solution can be found. Since the total amount of iodine originally formed is known the amount that reacts with the ascorbic acid is found by difference. Therefore the amount of ascorbic acid that reacts with this amount of iodine can be found.

Instructions

1. Set up the microscale titration apparatus (see p. 8).

2. Fill the apparatus with sodium thiosulphate solution (see p. 9).

3. Using the glass pipette add 2 cm³ of potassium iodate solution to the beaker.

4. Measure, using the measuring cylinder, 3 cm³ of potassium iodide solution, then add this to the beaker. (Note: the potassium iodide solution is added in slight excess.)

5. Add three drops of sulphuric acid. A yellow-brown colour appears due to iodine.

6. Add a few drops of starch solution. A deep blue-black colour forms.

7. Using the glass pipette add 1 cm³ of the fruit drink to the beaker and swirl gently.

THE ROYAL
SOCIETY OF
CHEMISTRY

8. Titrate the remaining iodine in the beaker against the sodium thiosulphate solution. (The beaker can be swirled very gently to mix the chemicals. Alternatively, the tip of a plastic pipette can be used as a mini stirring rod.) The disappearance of the deep blue-black colour marks the end-point.

9. Do a duplicate titration and check the agreement between the two titres. If it is acceptable take the mean value of the two titres and use it for your calculations.

Calculations

A specimen result and calculation is given below. Study this carefully and use it as a guide for working out the vitamin C content of your fruit drink.

The volume of thiosulphate delivered during the titration = 0.74 cm^3.

The concentration of thiosulphate = 0.010 mol dm^{-3}.

Therefore the number of moles of thiosulphate =

$$\frac{0.74 \times 0.010}{1000} = 7.4 \times 10^{-6}$$

Therefore the number of moles of iodine that this reacts with during the titration is 3.7×10^{-6}

The total number of moles of iodine produced in the reaction between iodate, iodide and sulphuric acid based on using 2 cm^3 of iodate with a concentration of 0.0012 mol dm^{-3} =

$$\frac{3 \times 2 \times 0.0012}{1000} = 7.2 \times 10^{-6}$$

Therefore the number of moles of iodine that reacts with the ascorbic acid is $7.2 \times 10^{-6} - 3.7 \times 10^{-6} = 3.5 \times 10^{-6}$.

Since 1 mole of iodine reacts with 1 mole of ascorbic acid then the number of moles of ascorbic acid is also 3.5×10^{-6}.

The volume of the fruit juice used is 1 cm^3. Therefore the number of moles of ascorbic acid in 1000 cm^3 = 3.5×10^{-3}.

The relative molar mass of ascorbic acid = 174.12 g. Therefore the mass of ascorbic acid (in 1000 cm^3) = 174.12 x 3.5×10^{-3} = 0.609 g.

Therefore the vitamin C content of the fruit drink = 61mg per 100 cm^3.

THE ROYAL
SOCIETY OF
CHEMISTRY

37. Using a microscale conductivity meter

In this experiment you will be using a conductivity meter (see p. 12) to test which solids and solutions/liquids conduct electricity.

Instructions

1. Cover the worksheet with a clear plastic sheet.

2. Add three drops of each of the solutions to the circles indicated below.

3. Place a small amount of each of the solids in the circle indicated below.

4. Test for conductivity by carefully placing just the tip of the electrodes in each of the substances in turn.

5. Make a table of your results.

6. Give explanations for your results trying to link the conductivity of a substance with its structure.

Copper sulphate solution   Sodium chloride solution   Sugar solution

Tap water   Deionised water   Copper foil

Iron nail   Aluminum foil   Pencil 'lead'

THE ROYAL
SOCIETY OF
CHEMISTRY

38. Electrolysis using a microscale Hoffman apparatus

In this experiment you will be investigating the electrolysis of sodium sulphate solution using a microscale Hoffman apparatus.

Instructions

1.   Set up the Hoffman apparatus in a clamp (see p. 15).

2.   Pour *ca* 40 cm³ of the sodium sulphate solution into a beaker and add a few drops of bromothymol blue indicator. Note the colour of the solution.

3.   Using a pipette carefully fill the electrolysis apparatus with the sodium sulphate solution.

4.   Plug the tops of each stem with a small piece of Blu-Tack®.

5.   Carefully attach the crocodile clips to the electrodes and record all your observations over the next 15 min.

6.   Disconnect the leads and try to give explanations for your observations.

Question

1.   Can you think of a way of testing for either of the gases that you have collected?

39. The determination of copper in brass

Version A

In this experiment you will be finding out how much copper there is in brass (an alloy of copper and zinc). You will dissolve the brass in nitric acid and compare the colour of the solution with that of solutions of various concentrations of copper.

Instructions

Part A   Preparing the brass solution

1.   Weigh out, accurately, about 0.3 g of brass in a 10 cm³ beaker.

2.   Put the beaker in a fume cupboard.

3.   Add 10 drops of nitric acid.

4.   When the reaction subsides add a further 10 drops of nitric acid.

5.   Repeat until all the brass has dissolved.

6.   Using the pipette, transfer the solution to a 10 cm³ volumetric flask. Add drops of water to the beaker to rinse and then transfer the washings to the flask. Make the volume in the flask up to the line with more water. Stopper the flask and then invert it a few times to mix.

Part B   Preparing the standard copper solutions

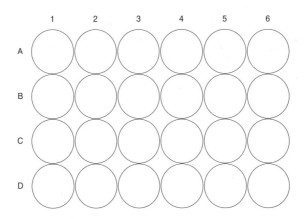

1.   Fill the well plate with solutions as indicated in the table below.

Well No	A1	A2	A3	A4	A5	A6	C1	C2	C3	C4	C5	C6
Drops of 0.50 mol dm⁻³ copper nitrate solution	8	10	12	14	16	18	20	22	24	26	28	30
Drops of water	32	30	28	26	24	22	20	18	16	14	12	10

THE ROYAL
SOCIETY OF
CHEMISTRY

There should be a total of 40 drops in each well.

Add 40 drops of the brass solution to well B3 (see diagram). Compare the intensity of the colour of your brass solution with the wells around it. The well that matches the intensity of colour of your brass solution represents the copper concentration in your brass solution – *eg* if well A6 matches the colour of your brass solution then the copper concentration will be $0.50 \times 18/40$ mol dm^{-3}.

Calculations

1.   Calculate the number of moles of copper in 10 cm^3 (the volume of the brass solution).

2.   Multiply the value you obtained in (1) by the relative atomic mass of copper (63.5) to give the mass of copper in the brass solution.

3.   Divide by the mass of brass used and express the result as a percentage.

Questions

1.   Does the zinc interfere in any way in this analysis? Give reasons for your answer.

2.   Can you suggest any way to improve the accuracy of this experiment?

Version B

In this experiment you will be finding out how much copper there is in brass (an alloy of copper and zinc). To do this you will dissolve the brass in nitric acid and compare the colour of the solution with that of solutions of various concentrations of copper.

Instructions

Part A   Preparing the brass solution

1.   Weigh out, accurately, about 0.3 g of brass in a 10 cm^3 beaker.

2.   Put the beaker in a fume cupboard.

3.   Add 10 drops of nitric acid.

4.   When the reaction subsides add a further 10 drops of nitric acid.

5.   Repeat until all the brass has dissolved.

6.   Using a pipette, transfer the solution to the 10 cm^3 volumetric flask. Add drops of water to the beaker to rinse and then transfer the washings to the flask. Make the volume in the flask up to the line with more water. Stopper the flask and then invert it a few times to mix.

THE ROYAL
SOCIETY OF
CHEMISTRY

Part B   Preparing the standard copper solutions

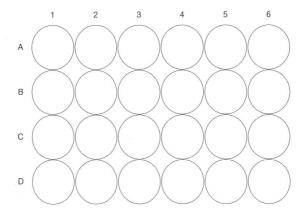

1.   Fill the well plate with the solutions as indicated in the table below.

Well No	A1	A2	A3	A4	A5	A6	C1	C2	C3	C4	C5	C6
Drops of 0.50 mol dm^{-3} copper nitrate solution	10	12	14	16	18	20	22	24	26	28	30	32
Drops of water	30	28	26	24	22	20	18	16	14	12	10	8

There should be a total of 40 drops in each well.

Add 40 drops of your brass solution to well B3 (see diagram). Compare the intensity of the colour of your brass solution with the wells around it.

From your results, calculate the copper content of your brass expressing your answer as a percentage.

Questions

1.   Does the zinc interfere in any way in this analysis? Give reasons for your answer.

2.   Can you suggest any way to improve the accuracy of this experiment?

THE ROYAL
SOCIETY OF
CHEMISTRY

40. Observing the lowering of a melting point

Pure organic compounds have characteristic sharp melting points. Any impurities lower the melting point and broaden its range. In this experiment you will mix two solid organic compounds with low melting points and observe the changes on mixing.

Instructions

1. Place a few crystals of hydroxybenzene (phenol) (mp 41 °C) on one side of a 5.5 cm plastic petri dish. (You should use the special procedure for sampling the bottle of hydroxybenzene. Your teacher will give you the instructions for this.)

2. On the other side of the dish place a few crystals of menthol (mp 28–30 °C).

3. Using the tip of a plastic pipette, mix the crystals in the centre of the dish and place the lid on the petri dish.

4. Observe for the next few minutes and try to give explanations for your observations.

At the end of the experiment, add sodium hydroxide solution dropwise from a plastic pipette. This will dissolve the hydroxybenzene. You should then wipe up the contents of the dish with tissue paper.

Hydroxybenzene
(phenol)

Menthol

THE ROYAL
SOCIETY OF
CHEMISTRY

41. Properties of stereoisomers

(R) - (+) - Limonene

(S) - (-) - Limonene

Limonene is a terpene present in orange and lemon oils. The enantiomers of limonene have markedly different odours – unscrew the tops of the bottles and sniff.

Questions

1. Would you expect the two stereoisomers of limonene to behave differently in their:

    a. reaction with bromine;

    b. reduction with hydrogen;

    c. melting point;

    d. boiling point;

    e. infrared spectrum;

    f. effect on plane of polarisation of plane-polarised light;

    g. combustion; and

    h. mass spectrum.

    Explain your answers.

THE ROYAL
SOCIETY OF
CHEMISTRY

42. Properties of the carvones

The two enantiomers of carvone have different odours – unscrew the tops of the bottles and sniff.

(R) - (-) - Carvone   (S) - (+) - Carvone

Questions

1. What are the similarities and differences in the structural formulae of the limonenes and the carvones?

2. Would you expect the two stereoisomers of carvone to behave differently in their:

    a. reaction with bromine;

    b. reduction with hydrogen;

    c. melting point;

    d. boiling point;

    e. infrared spectrum;

    f. effect on plane of polarisation of plane-polarised light;

    g. combustion; and

    h. mass spectrum.

    Explain your answers.

THE ROYAL
SOCIETY OF
CHEMISTRY

43. The formation of 2,4,6-trichlorohydroxybenzene by the reaction between hydroxybenzene and chlorine gas

In this experiment you will be generating chlorine gas inside a plastic petri dish and reacting it with crystals of hydroxybenzene (phenol). You will detect the product by its distinctive smell.

Instructions

1.   Cover the worksheet with a clear plastic sheet.

2.   Place the base of the petri dish over the circle overleaf.

3.   Cut the end off the plastic pipette and place the small cup – the reaction vessel – at the edge of the petri dish as indicated.

4.   Using the hydroxybenzene (phenol) sampling technique (ask your teacher) place a small quantity of hydroxybenzene (phenol) in the petri dish as indicated.

5.   Add two drops of bleach to the reaction vessel followed by two drops of hydrochloric acid. Quickly place the lid on the petri dish.

6.   Leave for 15 min then take off the lid. What kind of smell do you recognise?

7.   When you have finished add a few drops of sodium hydroxide to the reaction mixture to dissolve the solid and then mop up the solution with tissues.

Questions

1.   What do you observe and can you write equations for the reactions occurring:

   a.   to produce chlorine; and

   b.   how the chlorine reacts with the hydroxybenzene. What type of reaction is this?

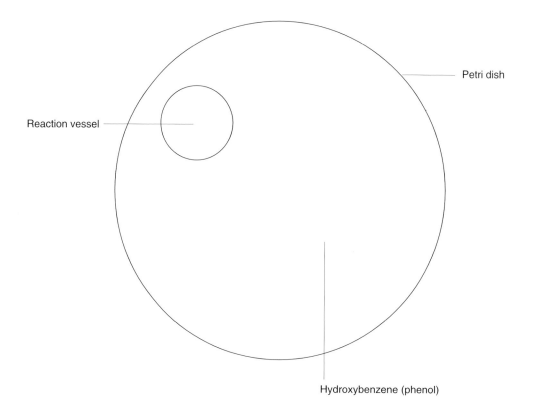

Petri dish

Reaction vessel

Hydroxybenzene (phenol)

THE ROYAL
SOCIETY OF
CHEMISTRY

44. The chemical properties of hydroxybenzene

In this experiment you will be observing and interpreting some of the chemical reactions of hydroxybenzene (phenol) inside a plastic petri dish.

Instructions

1. Cover the worksheet with a clear plastic sheet.

2. Place the base of the petri dish over the circle below. Using the hydroxybenzene (phenol) sampling procedure (ask your teacher) place small quantities of hydroxybenzene (phenol) in the petri dish over each of the five small circles.

3. In circle 1 add two drops of water, leave for 1 minute then add one drop of full-range indicator solution.

4. In circle 2 add two drops of 1 mol dm^{-3} nitric acid. Observe any changes over the next 5 min.

5. In circle 3 add two drops of iron(III) solution.

6. In circle 4 add two drops of sodium carbonate solution. Bearing in mind any conclusions that you arrived at in observing circle 1 what might you be looking for here?

7. In circle 5 add two drops of sodium hydroxide solution. Observe closely over the next minute. Add one drop of hydrochloric acid and observe closely.

When you have finished add drops of sodium hydroxide to the petri dish to dissolve the hydroxybenzene (phenol) and then mop up with tissue.

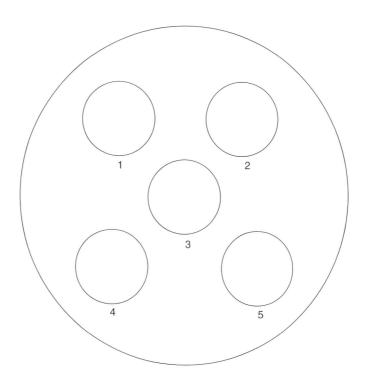

Questions

1. Can you give explanations for your observation?

2. Can you write equations for the reactions you have observed?

THE ROYAL
SOCIETY OF
CHEMISTRY

45. A test to distinguish between methanol and ethanol

These two important alcohols may be chemically distinguished by using the iodoform reaction. You will be doing this reaction in microscale using a well-plate.

Instructions

1. Add 10 drops of methanol to well A1 in your well-plate (see diagram).

2. Add 10 drops of ethanol to well A2.

3. Add 25 drops of iodine solution to each alcohol in the wells.

4. Add 10 drops of sodium hydroxide solution to each alcohol.

5. Gently swirl the well-plate a few times. The dark colour of the iodine should start to fade.

6. After 2 min carefully observe the two wells. What differences do you notice? Can you explain these differences?

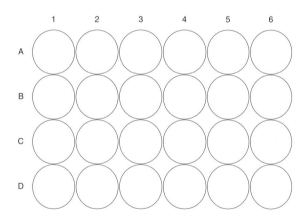

THE ROYAL
SOCIETY OF
CHEMISTRY

46. The formation of solid derivatives of aldehydes and ketones using 2,4-dinitrophenylhydrazine (Brady's Test)

In this experiment you will be adding various liquid aldehydes and ketones to 2,4-dinitrophenylhydrazine to form solid derivatives. To show that the reaction does not occur with alcohols you will also do the test with methanol and ethanol.

Instructions

1. In a well-plate, add 10 drops of 2,4-dinitrophenylhydrazine solution to each of the wells A1–A5 (see diagram).

2. Carefully add three drops of ethanal to well A1 (ethanal is very volatile!).

3. Repeat adding three drops of the other liquids to wells A2–A5.

4. Observe any changes over the next few minutes.

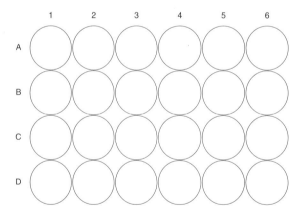

Questions

1. Can you write equations for any reactions occurring?

2. Why do you think that these reactions serve such a useful purpose in identifying aldehydes and ketones?

3. Aldehydes and ketones will also form derivatives with hydrazine itself. What is the purpose of using 2,4-dinitrophenylhydrazine in this experiment instead of hydrazine?

47. Testing for unsaturation using potassium manganate(VII)

In this experiment a solution of potassium manganate(VII) in propanone is used to detect whether an organic compound is unsaturated.

Instructions

1. Cut the end off a plastic pipette as shown below and place the cup in a beaker.

cut here

2. Carefully add a few crystals of potassium manganate(VII) to the cup.

3. Add propanone to the cup until it is about half-full. You will notice that the potassium manganate(VII) dissolves to give a purple solution.

    Is it surprising that potassium manganate(VII) dissolves in an organic solvent?

4. Cut the ends off three pipettes to make small reaction vessels as shown below and place them in the lid of a plastic petri dish.

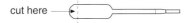

cut here

5. Using a plastic pipette, add four drops of the potassium manganate(VII) in propanone solution to each of the reaction vessels.

6. Put three drops of each of the organic liquids under test in the reaction vessel and observe any changes over the next few minutes.

7. Mop up the liquid with tissue paper when you have finished.

Question

1. Which types of organic liquids react with potassium manganate(VII)?

THE ROYAL
SOCIETY OF
CHEMISTRY

48. Preparing and testing ethyne

In this experiment you will be generating ethyne gas inside a plastic petri dish and testing its properties using a solution of potassium manganate(VII) in propanone.

Instructions

1. Cover the worksheet with a clear plastic sheet.

2. Place the base of the petri dish over the circle below.

2. Cut off the ends of two plastic pipettes (as shown below) and place them inside the petri dish.

cut here

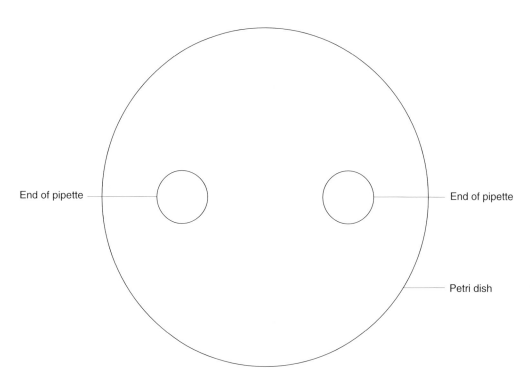
End of pipette   End of pipette

Petri dish

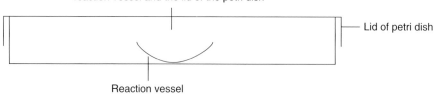
There must be a gap between the top of the reaction vessel and the lid of the petri dish

Lid of petri dish

Reaction vessel

THE ROYAL
SOCIETY OF
CHEMISTRY

3.   Cut off the bulb of a plastic pipette as shown below and place it in a beaker.

cut here

4.   Carefully add a few crystals of potassium manganate(VII) to the pipette.

5.   Add propanone to the pipette until it is about half-full.

6.   Using a pipette, add four drops of the potassium manganate(VII) in propanone solution to one of the pipette ends in the petri dish.

7.   Using tweezers carefully place one small lump of calcium carbide into the other pipette end.

8.   Carefully add four drops of deionised water to the calcium carbide and quickly place the lid on the petri dish.

9.   Observe any changes over the next few minutes.

10.   When no more gas is formed add one drop of full-range indicator solution to the residue of the calcium carbide and observe.

THE ROYAL
SOCIETY OF
CHEMISTRY

49. Testing for unsaturation using bromine

Preparing a solution of bromine in hexane

Elemental bromine is formed by a reverse disproportionation reaction between bromate ions, bromide ions and acid:

$$BrO_3^- (aq) + 5Br^-(aq) + 6H^+(aq) \rightarrow 3Br_2(aq) + 3H_2O(l)$$

The bromine is extracted into hexane in a plastic pipette which serves as a separating funnel. The resulting solution is decanted into a well-plate chamber and can then be used to test for unsaturation in organic compounds.

Instructions

1.   Put 10 drops of potassium bromate(V) solution into one well in a well-plate.

2.   Add 20 drops of potassium bromide solution.

3.   Add five drops of hydrochloric acid.

4.   Leave for 5 min to allow the bromine to form fully.

5.   Add hexane to the well until it is about half-full.

6.   Using your plastic pipette, take up all the liquid in the well and invert the pipette. You should see two layers – the lower (aqueous) layer which should be coloured red-yellow by the bromine and the upper layer which should be colourless. Notice, too, the shape of the meniscus at the interface.

7.   Gently flick the bulb of the pipette. This will mix the liquids and allow the bromine to be extracted into the upper hexane layer. Why?

8.   When the upper layer is coloured red-yellow and the lower layer is colourless your extraction is complete.

9.   Very carefully invert the pipette again and decant the lower aqueous layer into a well in your well-plate.

10.   Into another well decant the upper layer of bromine dissolved in hexane.

      This is the solution you will use for carrying out tests for unsaturation.

Testing for unsaturation using bromine

The solution of bromine in hexane is used to detect whether an organic compound is unsaturated. The solution easily mixes with non-polar organic compounds such as cyclohexane, cyclohexene and limonene.

Instructions

1.   Using a plastic pipette, add three drops of the bromine solution to each of the three wells in the well-plate.

2.   Put three drops of each of the organic liquids under test in the wells and observe any changes.

Question

1.   Which types of liquid decolourise bromide?

THE ROYAL
SOCIETY OF
CHEMISTRY

50. The oxidation of alcohols

In this experiment you will be testing various alcohols to see whether they can be oxidised by a solution of acidified potassium dichromate.

Instructions

1. Put 10 drops of acidified potassium dichromate solution into each of the wells A1–A5 and B3 (see diagram).

2. Add two drops of alcohol to the wells as follows:

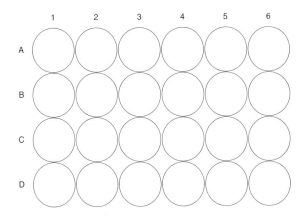

Well No	Alcohol
A1	Methanol
A2	Ethanol
A3	Propan-1-ol
A4	Propan-2-ol
A5	2-Methylpropan-2-ol

Do not put any alcohol into well B3 – this well is used as a control.

3. Observe the wells over the next 15 minutes and record any changes you see.

Questions

1. How do you explain any colour changes you see?

2. Which alcohols have been oxidised?

3. Can you find a connection between the ease of oxidation of an alcohol and its structure?

THE ROYAL
SOCIETY OF
CHEMISTRY

51. The oxidation of cyclohexanol by potassium dichromate

In this experiment you will be observing the effect of oxidising a secondary alcohol using an acidified solution of potassium dichromate.

When you are trying to interpret your observations bear in mind that when a substance is oxidised another substance must undergo a simultaneous reduction. Consider also any differences in the physical properties of the products compared to the reactants (you will need a data book to help you here).

Instructions

1.   Cover the worksheet with a clear plastic sheet.

2.   Place the base of the petri dish over the circle below.

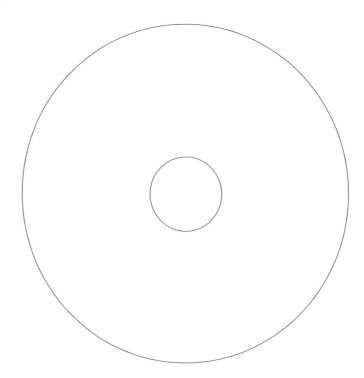

3.   Add 10 drops of the acidified potassium dichromate solution in the centre of the plastic petri dish (within the small circle).

4.   Add one drop of cyclohexanol.

5.   Observe over the next few minutes.

Question

1.   What do you observe? Give explanations for your observations.

THE ROYAL
SOCIETY OF
CHEMISTRY

52. The oxidation of cyclohexanol by nitric acid

In this experiment you will be oxidising cyclohexanol using nitric acid. In this reaction the nitric acid breaks open the six-carbon ring to form the dicarboxylic acid, 1,6-hexanedioic acid (adipic acid). Whereas cyclohexanol is a liquid, 1,6-hexanedioic acid (adipic acid) is a solid and you can measure its melting point.

The reaction is:

OH

[O] →

CO_2H
|
$(CH_2)_4$
|
CO_2H

Cyclohexanol

1,6-hexanedioic acid
(Adipic acid)

Instructions

1. Half-fill a 100 cm^3 beaker with deionised water, and heat to 80–90 °C.

2. Add 1 cm^3 of nitric acid to a test-tube and place in the water bath.

3. Carefully add six drops of cyclohexanol to the test-tube. You will notice some bubbling and the nitric acid turns brown.

4. Leave for 10 min.

5. Remove the test-tube from the water bath and allow to cool to room temperature.

6. Cool further in an ice bath – crystals should form.

7. Filter off the crystals, wash with 2 cm^3 of deionised water and dry them.

8. Measure the melting point of your product.

Question

1. What is the melting point of your product? How does it compare with the value from data books? Can you explain any variations?

THE ROYAL
SOCIETY OF
CHEMISTRY

53. The microscale synthesis of aspirin

In this experiment you will be preparing 2-ethanoyloxybenzenecarboxylic acid (aspirin) from the reaction between 2-hydroxybenzoic acid (salicylic acid) and ethanoic anhydride.

The reaction is:

2-Hydroxybenzoic acid   Ethanoic anhydride   2-ethanoyloxybenzenecarboxylic acid
(Salicylic acid)   (Aspirin)

Instructions

1.   Half-fill a 50 cm³ beaker with deionised water, and heat to 70–80°C.

2.   Weigh 0.23 g of 2-hydroxybenzoic acid (salicylic acid) into a test-tube.

3.   Add 25 drops of ethanoic anhydride followed by one drop of 85% phosphoric acid.

4.   Place in the water bath and leave for 15 min.

5.   While still warm add 1.5 cm³ of deionised water (use the measuring cylinder) and cool to room temperature until crystallisation begins, then cool in an ice bath.

6.   Filter through a small filter funnel and recrystallise in a test-tube using a mixture of 0.7 cm³ ethanol and 2 cm³ of deionised water.

THE ROYAL
SOCIETY OF
CHEMISTRY

54. The analysis of aspirin tablets

In this experiment you will be finding out how much 2-hydroxybenzoic acid (salicylic acid) is present in 2-ethanoyloxybenzenecarboxylic acid (aspirin) tablets.

2-Hydroxybenzoic acid (salicylic acid) is formed in the following reaction:

2-Ethanoyloxybenzenecarboxylic acid   2-Hydroxybenzoic acid
            (Aspirin)   (Salicylic acid)

Instructions

Part A   The preparation of standard solutions

In this part of the experiment you will be preparing a set of standard solutions with different colour intensities from the standard 2-hydroxybenzoic acid (salicylic acid) solution. You will be using these to match the intensity of the colour produced from the 2-ethanoyloxybenzenecarboxylic acid (aspirin) solution and so find out how much 2-hydroxybenzoic acid (salicyclic acid) there is in your 2-ethanoyloxybenzenecarboxylic acid (aspirin) tablet.

Taking your 24-well plate, add drops of solutions as indicated below:

Well no	A1	A2	A3	A4	A5	A6
No. of drops of: 2-ethanoyloxybenzene-carboxylic acid (salicyclic acid) soln.	5	15	25	35	45	50
Water	45	35	25	15	5	0
Iron(III) nitrate solution	5	5	5	5	5	5
Resulting mass (mg) of 2-hydroxybenzoic acid (salicyclic acid) per 25 cm^3 solution	0.25	0.75	1.25	1.75	2.25	2.5

Part B   The analysis of 2-ethanoyloxybenzenecarboxylic acid (aspirin) tablets

1.   Record the mass of a 2-ethanoyloxybenzenecarboxylic acid (aspirin) tablet and place it in a 100 cm^3 beaker.

2.   Add 10 cm^3 of the 50% ethanol–water mixture (from a measuring cylinder) and swirl the mixture. The tablet will begin to disintegrate.

3.   Using the microscale filtration method (p. 5), filter the mixture into a 25 cm^3 volumetric flask. Wash the beaker with a small quantity of the ethanol–water mixture and add to the flask. Make up to the mark, stopper and mix.

4. Add 50 drops of this 2-ethanoyloxybenzenecarboxylic acid (aspirin) solution to well B3 followed by five drops of the iron(III) nitrate solution.

5. Match the colour to that of one of the standard solutions.

Calculations

Calculate the percentage of 2-hydroxybenzoic acid (salicylic acid) in the 2-ethanoyloxybenzenecarboxylic acid (aspirin) tablet as follows.

1. Identify the standard well that matches the colour intensity of the 2-ethanoyloxybenzenecarboxylic acid (aspirin) sample well.

2. The mass of 2-hydroxybenzoic acid (salicylic acid) (in 25 cm³) in the solution from this standard well is therefore the same as the mass of 2-hydroxybenzoic acid (salicylic acid) in the 25 cm³ of solution of your 2-ethanoyloxybenzenecarboxylic acid (aspirin) tablet solution.

3. Divide this mass (mg) by the mass of your 2-ethanoyloxybenzenecarboxylic acid (aspirin) tablet (mg) and multiply this value by 100 to give a percentage by mass.

Question

1. By considering the equation for the formation of 2-hydroxybenzoic acid (salicylic acid) from 2-ethanoyloxybenzenecarboxylic acid (aspirin), are there any differences in how much 2-hydroxybenzoic acid (salicylic acid) is present in both old and new bottles of 2-ethanoyloxybenzenecarboxylic acid (aspirin) tablets?

THE ROYAL
SOCIETY OF
CHEMISTRY

55. The conversion of alcohols to halogenoalkanes

In this experiment you will be looking at an alternative method for replacing a hydroxy group by a halogen using phosphorus tribromide (PBr$_3$).

Instructions

This experiment must be done in a fume cupboard. You should wear disposable gloves when using PBr$_3$.

1. Put 20 drops of ethanol into the test-tube.

2. By adding one drop at a time, put 15 drops of phosphorus tribromide in the test-tube.

3. Leave the mixture to stand for 10 min.

4. Put 2 cm^3 of deionised water in a measuring cylinder and, one drop at a time, add the water to the test-tube.

5. Observe the test-tube carefully. What do you notice?

THE ROYAL
SOCIETY OF
CHEMISTRY

56. The microscale synthesis of azo dyes

In this experiment you will be synthesing an azo dye and using it to dye a piece of cotton.

The reactions are:

1. Diazotisation

NH_2 + HCl / NaNO$_2$, <5 °C → $N^+\equiv N$ Cl$^-$

Aminobenzene

2. Coupling

$N^+\equiv N$ Cl$^-$ + 2-Naphthol (OH) → Azo dye

Instructions

1. Put eight drops of aminobenzene in a 10 cm³ beaker and add 30 drops of deionised water followed by 15 drops of concentrated hydrochloric acid. Swirl the beaker and then put it in an ice bath.

2. Weigh 0.15 g of sodium nitrite into another beaker and add 1 cm³ of deionised water. Cool the beaker in the ice bath. Add one spatula of urea (this prevents side reactions occurring).

3. Mix the contents of the two beakers together and keep in the ice bath.

4. Weigh 0.45 g of 2-naphthol into another beaker and add 3 cm³ of sodium hydroxide solution. Swirl to dissolve.

5. Take a piece of cotton cloth 2 x 2 cm² and, using tweezers, dip it into the 2-naphthol solution. Allow the solution to completely soak the cotton.

6. Dip the cloth completely into the diazonium salt solution. A red dye forms in the fibres, dyeing the cloth.

7. Take the cloth out, wash it under the tap and leave to dry.

THE ROYAL
SOCIETY OF
CHEMISTRY

57. The microscale synthesis of indigo dye

In this experiment you will be doing a microscale synthesis of the dye indigo. This dye has been known for a very long time (in Britain it was being used before the Romans first came to the country) and used to be obtained naturally from woad. It was first produced synthetically by the German chemist Baeyer in 1879 and this work formed the foundation of much of the modern organic chemical industry. You will probably be very familiar with indigo – it is the colouring in blue jeans!

Instructions

1. Weigh out approximately 0.1 g of 2-nitrobenzaldehyde into a test-tube.

2. Add 2 cm³ of propanone (from a measuring cylinder) and swirl gently to dissolve the solid.

3. Add 25 drops of deionised water and swirl gently.

4. Slowly add 20 drops of sodium hydroxide solution to the solution. The solution quickly darkens and a purple solid (indigo) should precipitate out.

5. Leave for 5 min to complete the precipitation.

6. Filter the solution washing with deionised water until the washings are colourless, and set aside to dry.

7. Describe the appearance of your product.

The structure of indigo is shown below.

Indigo

58. The preparation of ethyl benzoate

In this experiment you will be preparing the ester ethyl benzoate by warming ethanol and benzoic acid in a plastic pipette in a water bath.

Instructions

Appropriate care must be taken when using concentrated acids.

1. Weigh 0.24 g of benzoic acid into a 10 cm³ beaker.

2. Add 1 cm³ of ethanol and swirl to dissolve.

3. Transfer, using a plastic pipette, to a shortened standard form pipette.

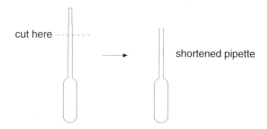

4. Add one drop of concentrated sulphuric acid.

5. Cut the bulb off another pipette and place it over the shortened pipette end as shown:

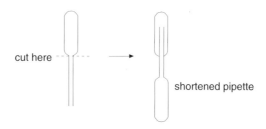

6. Place in a water bath at 70–80 °C for 10 min.

7. Remove from the water bath, take off the top and waft the scent towards your nose.

Questions

1. How would you describe the smell?

2. Write an equation for the reaction.

THE ROYAL
SOCIETY OF
CHEMISTRY

59.   The treatment of oil spills

In this experiment you will be looking at an unusual and interesting way of chemically treating a microsize oil spill.

Instructions

1. Half-fill a 100 cm^3 beaker with water.

2. Using your pipette, add some oil or paraffin to the beaker to give a thin layer on top of the water.

3. Cut off the end of a pipette to form a scoop as shown below.

cut here

4. Add two scoops of polymer powder to the beaker and stir with the end of the pipette.

Questions

1. What do you observe?

2. How do you explain your observations?

3. If you were to do this experiment on a large scale to try to deal with an oil slick at sea what would be the advantages of using this polymer powder and what difficulties might you encounter?

THE ROYAL
SOCIETY OF
CHEMISTRY

60. Detecting starch in food

In this experiment you will be testing various foods to see whether they contain starch.

Instructions

1. Cover the worksheet with a clear plastic sheet.

2. Place a small quantity of each food in the circles below.

3. Using tweezers, carefully put one crystal of potassium iodide on each piece of food.

4. Carefully add one drop of bleach to each and observe any changes.

Questions

1. Can you offer an explanation for your observations?

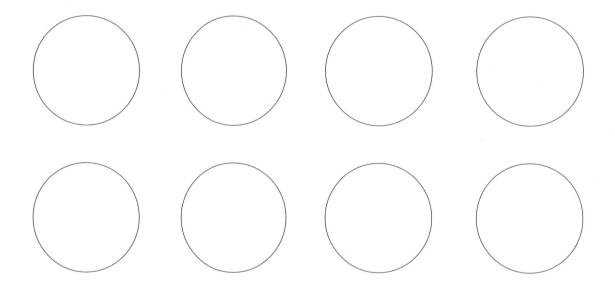

Teacher's guide

THE ROYAL
SOCIETY OF
CHEMISTRY

1.   Acids and bases

Topic

Acids and bases, indicators.

Level

Primary and early secondary.

Timing

15 min.

Apparatus (per group)

- ▼ One student worksheet
- ▼ One clear plastic sheet (*eg* ohp sheet)
- ▼ Plastic pipettes.

Chemicals (per group)

Solutions contained in plastic pipettes, see p. 2

▼	Sodium hydroxide	1 mol dm^{-3}
▼	Hydrochloric acid	1 mol dm^{-3}
▼	Sulphuric acid	1 mol dm^{-3}
▼	Nitric acid	1 mol dm^{-3}
▼	Sodium carbonate	0.5 mol dm^{-3}
▼	Ammonia solution	1 mol dm^{-3}
▼	Vinegar	
▼	Lemon juice	
▼	Household bleach (diluted 1:1 with water)	
▼	Soap solution	
▼	Full-range indicator solution (diluted 1:1 with deionised water).	

(Full-range indicator is a solution in propanol which has a low surface tension and spreads out if used neat. Adding water increases the surface tension while still keeping the indicator in solution.)

Safety

Students must wear eye protection.
   It is the responsibility of the teacher to carry out a risk assessment.

THE ROYAL
SOCIETY OF
CHEMISTRY

2.   A chemical reaction

Topic

Chemical reactions.

Level

Pre-16.

Timing

5 min.

Apparatus (per group)

- ▼   One student worksheet
- ▼   One clear plastic sheet (*eg* ohp sheet)
- ▼   Tweezers
- ▼   One drinking straw
- ▼   Magnifying glass
- ▼   One plastic pipette.

Chemicals (per group)

- ▼   Lead nitrate crystals
- ▼   Potassium iodide crystals
- ▼   Deionised water (in plastic pipette).

Observations

No change is observed when the lead nitrate crystals are mixed with the potassium iodide powder. However, the two compounds react immediately when dissolved in water, to form a yellow precipitate of lead iodide.

The experiment provides several useful points for discussion about chemical reactions. Two solids do not usually react together since the particles in each are held rigidly in the lattice. When separated on dissolution, however, they interact readily.

Safety

Students must wear eye protection.

It is the responsibility of the teacher to carry out a risk assessment.

THE ROYAL
SOCIETY OF
CHEMISTRY

3.   Observing chemical changes

Topic

Displacement, redox and precipitation reactions. Chemistry and colour.

Level

Any.

Timing

20 min.

Apparatus (per group)

- ▼   One student worksheet
- ▼   One clear plastic sheet (*eg* ohp sheet)
- ▼   Magnifying glass.

Chemicals (per group)

Solutions contained in plastic pipettes, see p. 2

▼	Barium nitrate	$0.2 \ mol \ dm^{-3}$
▼	Sodium sulphate	$0.5 \ mol \ dm^{-3}$
▼	Lead nitrate	$0.5 \ mol \ dm^{-3}$
▼	Ammonia solution	$3 \ mol \ dm^{-3}$
▼	Ammonium vanadate(V)	$0.2 \ mol \ dm^{-3}$ (acidified with sulphuric acid)
▼	Hydrochloric acid	$1 \ mol \ dm^{-3}$
▼	Sodium hydroxide	$1 \ mol \ dm^{-3}$
▼	Potassium manganate(VII)	$0.01 \ mol \ dm^{-3}$
▼	Silver nitrate	$0.1 \ mol \ dm^{-3}$
▼	Copper(II) sulphate	$0.2 \ mol \ dm^{-3}$
▼	Iron(II) sulphate	$0.2 \ mol \ dm^{-3}$
▼	Iron(III) nitrate	$0.2 \ mol \ dm^{-3}$
▼	Potassium thiocyanate	$0.2 \ mol \ dm^{-3}$
▼	Zinc metal granules.	

Observations

1.   A dense white precipitate of barium sulphate forms. Barium sulphate is used as a barium meal in medicine since it is opaque to X-rays. Because it is very insoluble it is non-toxic, unlike other, soluble, barium compounds.

2.   A bright yellow precipitate of lead nitrate forms. Lead nitrate is a very effective pigment but it is toxic.

THE ROYAL
SOCIETY OF
CHEMISTRY

3. A deep red colour is produced due to iron(III) thiocyanate ions . This reaction is used to test for the presence of iron.

4. A deep blue colour of tetra-amminocopper(II) forms. There may also be some light blue precipitate of copper(II) hydroxide.

5. Bubbles (of hydrogen) are seen. The yellow colour of the ammonium vanadate gradually changes (as the vanadium is reduced) to blue owing to the formation of the vanadium(IV) ion (VO^{2+}). The colour then changes to green due to the vanadium(III) ion (V^{3+}) and finally to lilac due to the vanadium(II) ion (V^{2+}). The changes in oxidation states of vanadium salts have been investigated for applications in battery technology.

6. A greenish precipitate of iron(II) hydroxide forms. This gradually changes to the brown iron(III) hydroxide as the iron is oxidised.

7. The deep purple colour of the potassium manganate(VII) gradually fades first to the brown manganese(IV) dioxide and then to the pale pink manganese(II) ion (Mn^{2+}). Manganese(II) compounds in solution usually appear virtually colourless. However, a solid manganese(II) salt is pink.

8. Barium hydroxide forms. This is soluble so nothing is seen at first. Barium hydroxide is alkaline and gradually absorbs carbon dioxide from the air to form the insoluble barium carbonate. The drop takes on a hazy appearance as a skin of barium carbonate forms on the surface.

9. A glittering of metallic silver forms as the iron(III) reduces the silver nitrate. This is seen clearly using a magnifying glass.

10. The surfaces of the pieces of zinc turn red-brown as copper metal deposits via a displacement reaction. The blue colour of the copper(II) sulphate solution fades.

Note

Both procedure 9 and procedure 10 involve the displacement of a valuable, but less reactive, metal using a less valuable, but more reactive, metal. This could be used as a topic for discussion.

Safety

Students must wear eye protection.
It is the responsibility of the teacher to carry out a risk assessment.

THE ROYAL
SOCIETY OF
CHEMISTRY

4.   The reaction of metals with acids

Topic

Metals – reactions with acids; reactivity series.

Level

Pre-16.

Timing

20 min.

Description

In this experiment students observe the reactions between metals and acids.

Apparatus (per group)

▼   One student worksheet

▼   One clear plastic sheet (*eg* ohp sheet)

▼   Magnifying glass.

Chemicals (per group)

Solutions contained in plastic pipettes, see p. 2

▼	Hydrochloric acid	1 mol dm^{-3}
▼	Dilute nitric acid	1 mol dm^{-3}
▼	Concentrated nitric acid	5 mol dm^{-3}
▼	Sulphuric acid	1 mol dm^{-3}
▼	Magnesium ribbon	
▼	Zinc metal – small granules	
▼	Iron filings	
▼	Tin granules	
▼	Copper turnings.	

Observations

The magnesium ribbon reacts vigorously with each acid. The zinc and iron also react, but less vigorously. In each case hydrogen gas is produced as well as the metal salt. The reaction between iron and nitric acid eventually produces a red-brown rust colour (iron(III) oxide). Students could link this with corrosion and acid rain. Tin and copper do not react with the hydrochloric and sulphuric acids but a few bubbles may be seen (using the magnifying glass) with the nitric acid.

The copper reacts with the 5 mol dm^{-3} nitric acid to produce a blue solution and bubbles (of brown nitrogen dioxide).

Students can write word and symbol equations for these reactions.

Safety

Students must wear eye protection.

It is the responsibility of the teacher to carry out a risk assessment.

5. Displacement reactions of metals

Topic

Metals – reactions with acids; reactivity series.

Level

Pre-16.

Timing

20 min.

Description

In this experiment students observe the reactions between metals and metal salt solutions.

Apparatus (per group)

▼   One student worksheet

▼   One clear plastic sheet (*eg* ohp sheet)

▼   Magnifying glass.

Chemicals (per group)

Solutions contained in plastic pipettes, see p. 2

▼   Copper(II) sulphate   0.2 mol dm^{-3}

▼   Iron(III) nitrate   0.2 mol dm^{-3}

▼   Magnesium nitrate   0.2 mol dm^{-3}

▼   Zinc chloride   0.2 mol dm^{-3}

▼   Magnesium ribbon

▼   Zinc metal – small granules

▼   Iron filings or small nails

▼   Copper turnings.

Observations

The zinc granules and magnesium ribbon rapidly darken in copper sulphate solution as they become covered with a layer of copper. Iron also reacts but the change is not so clear. Magnesium and zinc react with the iron(III) nitrate, the solution gradually darkens.

No reaction occurs between magnesium sulphate and any of the metals. Students should observe no change between any of the metals and a salt solution of the same metal.

Safety

Students must wear eye protection.

It is the responsibility of the teacher to carry out a risk assessment.

THE ROYAL
SOCIETY OF
CHEMISTRY

6.   Redox reactions

Topic

Transition elements – redox reactions; electrochemistry – redox reactions – changes in reduction potentials down Group VII.

Level

Post-16.

Timing

20 min.

Apparatus (per group)

▼   One student worksheet

▼   One clear plastic sheet (*eg* ohp sheet)

▼   Magnifying glass.

Chemicals (per group)

Solutions contained in plastic pipettes, see p. 2

▼   Potassium bromide $0.2 \ mol \ dm^{-3}$

▼   Potassium iodide $0.2 \ mol \ dm^{-3}$

▼   Sodium chloride $0.2 \ mol \ dm^{-3}$

▼   Silver nitrate $0.1 \ mol \ dm^{-3}$

▼   Copper(II) sulphate $0.2 \ mol \ dm^{-3}$

▼   Iron(II) sulphate $0.2 \ mol \ dm^{-3}$

▼   Iron(III) nitrate $0.2 \ mol \ dm^{-3}$

▼   Potassium thiocyanate $0.1 \ mol \ dm^{-3}$

▼   Starch solution   (freshly made).

Observations

Part A

No changes are observed on adding chloride or bromide to the copper(II) solution. However, the addition of iodide gives an immediate light brown precipitate of copper(I) iodide. The addition of starch solution gives the intense blue-black colour characteristic of the starch–iodine complex (see reference). Iodide reduces copper(II):

$$2Cu^{2+}(aq) + 4I^-(aq) \rightarrow 2CuI(s) + I_2(s)$$

Part B

The addition of iron(II) solution to silver nitrate produces silver metal by reduction. Glittering can be seen in the drop.

THE ROYAL
SOCIETY OF
CHEMISTRY

The addition of a drop of thiocyanate produces a deep red colour indicative of iron(III). A whitish precipitate of silver thiocyanate can also be seen.

The second part of this experiment is for students to do sequential reactions of thiocyanate with silver(I), iron(II) and iron(III), helping them to interpret this redox reaction.

Note

Unless very pure and freshly prepared, iron(II) solutions will contain a small amount of iron(III) which gives a slight red coloration in the reaction between the iron(III) solution and the thiocyanate. However, the intensity of the colour is less than that observed in the reaction between iron(III) solution and thiocyanate ions. This point could be explored further in subsequent discussions on the purity of chemicals.

Reference

School Sci. Rev., 1990, **72**, 104.

Safety

Students must wear eye protection.

It is the responsibility of the teacher to carry out a risk assessment.

THE ROYAL
SOCIETY OF
CHEMISTRY

7. The Periodic Table – solubility of sulphates and carbonates of Groups 1 and 2

Topic

Periodic Table – Groups 1 and 2.

Level

Pre-16 and post-16.

Timing

20 min.

Apparatus (per group)

▼   One student worksheet

▼   One clear plastic sheet (*eg* ohp sheet).

Chemicals (per group)

Solutions contained in plastic pipettes, see p. 2

▼   Magnesium nitrate   0.5 mol dm^{-3}

▼   Calcium nitrate   0.5 mol dm^{-3}

▼   Strontium nitrate   0.5 mol dm^{-3}

▼   Barium nitrate   0.2 mol dm^{-3}

▼   Lithium bromide   1 mol dm^{-3}

▼   Sodium chloride   0.5 mol dm^{-3}

▼   Potassium bromide   0.2 mol dm^{-3}

▼   Sodium carbonate   0.5 mol dm^{-3}

▼   Sodium sulphate   0.5 mol dm^{-3}.

Observations

There should be no precipitates in Group 1, indicating that all Group 1 carbonates and sulphates are soluble.

For Group 2, magnesium sulphate is soluble while strontium and barium sulphates are insoluble. Calcium sulphate is particularly interesting because although it is only sparingly soluble its solubility is much higher than is expected from the solubility product. This is due to ion pairing of the calcium and sulphate ions in aqueous solution. No precipitate will be seen.

The concepts of solubility product and ion pairing may be too complex for most pre-16 students.

Tips

Students might think that the Group 1 part of this experiment is rather dull. However, they can be told that chemistry experiments that seem to produce no visual results may nevertheless still produce useful information!

Students should also observe that all the precipitates are white not coloured. The accompanying solubility data will be useful.

Solubility data

Solubility in grams per 100 cm^3 of water at 20 °C (except where indicated with a superscript).

	Carbonate	Hydroxide	Sulphate	Fluoride
Magnesium	0.0106	0.0009	73.8	0.0076[28]
Calcium	0.0014	0.185	0.209	0.0016[18]
Strontium	0.0011	0.41	0.0113	0.012[27]
Barium	0.002	5.6	0.00022	0.12[25]

(Source: *CRC handbook of chemistry and physics*, 74th edn. 1993–4.)

Safety

Students must wear eye protection.

It is the responsibility of the teacher to carry out a risk assessment.

THE ROYAL
SOCIETY OF
CHEMISTRY

8.   The Periodic Table – properties of Group 2 elements

Topic

Periodic Table – Group 2

Level

Pre-16 and post-16.

Timing

20 min.

Apparatus (per group)

▼   One student worksheet

▼   One clear plastic sheet (*eg* ohp sheet).

Chemicals (per group)

Solutions contained in plastic pipettes, see p. 2

▼   Magnesium nitrate   0.5 mol dm^{-3}

▼   Calcium nitrate   0.5 mol dm^{-3}

▼   Strontium nitrate   0.5 mol dm^{-3}

▼   Barium nitrate   0.2 mol dm^{-3}

▼   Sodium hydroxide   1 mol dm^{-3}

▼   Sodium fluoride   0.5 mol dm^{-3}

▼   Sodium chloride   0.5 mol dm^{-3}

▼   Potassium bromide   0.2 mol dm^{-3}

▼   Potassium iodide   0.2 mol dm^{-3}

▼   Sodium carbonate   0.5 mol dm^{-3}

▼   Sodium sulphate   0.5 mol dm^{-3}.

Observations

Magnesium

No precipitates should be seen. All the compounds are colourless and soluble at these concentrations.

Calcium

An immediate white cloudiness is seen with the carbonate ions. No precipitates are seen with chloride, bromide or iodide but a cloudiness is seen with fluoride (due to its high lattice energy CaF_2 is insoluble).

THE ROYAL
SOCIETY OF
CHEMISTRY

Calcium hydroxide

This is clear at first but when left for a few minutes the drops become hazy as calcium carbonate is formed by absorbing carbon dioxide from the air:

$$Ca(OH)_2 + CO_2 \rightarrow CaCO_3 + H_2O$$

Calcium sulphate also appears clear (the solubility product is not realised at these concentrations due, possibly, to ion-pairing).

Strontium

The sulphate and carbonate are insoluble and a white cloudiness is seen. For the sulphate, the precipitate forms slowly.

The halides are all soluble, except for the fluoride. The hydroxide is clear at first but becomes hazy – similar to calcium.

Barium

The sulphate and carbonate give immediate white precipitates.

The halides are soluble except for the fluoride. The hydroxide is (like calcium and strontium) clear at first, becoming hazy due to the formation of barium carbonate.

Safety

Students must wear eye protection.

It is the responsibility of the teacher to carry out a risk assessment.

THE ROYAL
SOCIETY OF
CHEMISTRY

9.   The Periodic Table –
changes down the Group 7 elements

Topic

Halogens.

Level

Post-16.

Timing

10 min.

Apparatus (per group)

▼   One student worksheet

▼   One clear plastic sheet (*eg* ohp sheet).

Chemicals (per group)

Solutions contained in plastic pipettes, see p. 2

▼   Silver nitrate 0.1 mol dm^{-3}

▼   Lithium bromide 1 mol dm^{-3}

▼   Calcium nitrate 0.5 mol dm^{-3}

▼   Sodium fluoride 0.5 mol dm^{-3}

▼   Sodium chloride 0.2 mol dm^{-3}

▼   Potassium bromide 0.2 mol dm^{-3}

▼   Potassium iodide 0.2 mol dm^{-3}.

Observations

Students should observe that silver fluoride is soluble whereas the other silver halides are insoluble. For calcium halides, the fluoride is insoluble while the other halides are soluble. With the lithium halides, the fluoride is insoluble but it is difficult to get it to precipitate even at high concentrations. (This may be due to the effect of ion-pairing in solution so that ion activity, sufficient to reach the solubility product, is not realised.) The other lithium halides are all very soluble.

The general conclusion is that fluorides behave rather differently from the other halides and the first element in any Group generally behaves anomalously. Follow up discussion on this point could focus on some of the reasons for this, such as ion size, ionisation energy and lattice energy.

Safety

Students must wear eye protection.

It is the responsibility of the teacher to carry out a risk assessment.

THE ROYAL
SOCIETY OF
CHEMISTRY

10. Lead compounds – precipitation reactions and pigments

Topic

Lead compounds – precipitation reactions and pigments.

Level

Pre-16 and post-16.

Timing

20 min.

Apparatus (per group)

▼   One student worksheet

▼   One clear plastic sheet (*eg* ohp sheet).

Chemicals (per group)

Solutions contained in plastic pipettes, see p. 2

▼   Sodium hydroxide   1 mol dm^{-3}

▼   Lead nitrate   0.5 mol dm^{-3}

▼   Potassium iodide   0.2 mol dm^{-3}

▼   Sodium chloride   0.5 mol dm^{-3}

▼   Potassium bromide   0.2 mol dm^{-3}

▼   Sodium carbonate   0.5 mol dm^{-3}

▼   Sodium sulphate   0.5 mol dm^{-3}

▼   Potassium chromate   0.2 mol dm^{-3}.

Observations

Part A

The addition of solutions of each of the anions produces precipitates, which indicates that in general lead compounds are insoluble. The iodide is an intense yellow colour, the chromate(VI) is also yellow and both could be used as pigments except for the fact that lead compounds are toxic.

Part B

The fact that lead forms insoluble compounds is used as a basis for indicating the presence of anions in water. The addition of deionised water to lead nitrate gives no cloudiness. However, with tap water a cloudiness gradually develops if the water is from a hard water area since carbonates, sulphates or hydrogencarbonates may be present. If you live in a soft water area there will probably be no cloudiness. (One solution is to simulate hard water conditions.)

Safety

Students must wear eye protection.
It is the responsibility of the teacher to carry out a risk assessment.

THE ROYAL
SOCIETY OF
CHEMISTRY

11. The chemistry of silver

Topic

Transition elements – complex formation, redox reactions.
Light sensitive compounds – photography.

Level

Pre-16 and post-16.

Timing

20 min.

Apparatus (per group)

▼   One student worksheet

▼   One clear plastic sheet (*eg* ohp sheet)

▼   Piece of black or dark card

▼   Scissors

▼   Magnifying glass.

Chemicals (per group)

Solutions contained in plastic pipettes, see p. 2

▼   Ammonia solution   3 mol dm^{-3}

▼   Iron(II) sulphate   0.2 mol dm^{-3}

▼   Potassium bromide   0.2 mol dm^{-3}

▼   Potassium iodide   0.2 mol dm^{-3}

▼   Silver nitrate   0.1 mol dm^{-3}

▼   Sodium chloride   0.2 mol dm^{-3}.

Observations

The addition of chloride, bromide and iodide solutions to silver nitrate solution produces immediate precipitates – white, pale yellow and yellow respectively.

The appearance of the silver halide precipitates is retained if they are covered from the light. The colours of the silver chloride and bromide gradually darken due to the formation of grey metallic silver, while the iodide appears to be more stable when left exposed to light (artificial or sunlight).

Silver chloride dissolves readily in ammonia solution while the bromide partially dissolves and the iodide does not.

Adding iron(II) solution to the silver nitrate solution produces a glittering of metallic silver which can be seen using a magnifying glass.
The reaction is:

$$Ag^+ (aq) + Fe^{2+}(aq) \rightarrow Ag(s) + Fe^{3+}(aq)$$

Discussion of this reaction could centre around the respective redox potentials. The relevant standard redox potentials are:

THE ROYAL
SOCIETY OF
CHEMISTRY

$Ag^+ + e^- \rightarrow Ag \qquad E^\oplus = +0.80$ V

$Fe^{3+} + e^- \rightarrow Fe^{2+} \quad E^\oplus = +0.77$ V

Thus $E^\oplus = -0.03$ V for the reaction

As $\Delta G^\oplus = -zFE^\oplus$ thus $\Delta G^\oplus$ is +ve

Reference

F.A. Cotton and G. Wilkinson, *Advanced inorganic chemistry*, 5th edn, p 942.
London: John Wiley, 1988.

Safety

Students must wear eye protection.
    It is the responsibility of the teacher to carry out a risk assessment.

THE ROYAL
SOCIETY OF
CHEMISTRY

12. Iron chemistry
– variable oxidation state

Topic

Iron chemistry, transition elements, oxidation states, precipitation and redox reactions, complexes.

Level

Post-16.

Timing

20 min.

Apparatus (per group)

▼   One student worksheet

▼   One clear plastic sheet (*eg* ohp sheet)

▼   Magnifying glass.

Chemicals (per group)

Solutions contained in plastic pipettes, see p. 2

▼	Sodium hydroxide	1 mol dm^{-3}
▼	Potassium manganate(VII)	0.01 mol dm^{-3}
▼	Potassium iodide	0.2 mol dm^{-3}
▼	Iron(II) sulphate	0.2 mol dm^{-3}
▼	Iron(III) nitrate	0.2 mol dm^{-3}
▼	Silver nitrate	0.2 mol dm^{-3}
▼	Potassium thiocyanate	0.1 mol dm^{-3}
▼	Starch solution	(freshly made).

Observations

1. The addition of sodium hydroxide produces a gelatinous green precipitate with iron(II) solution and a brown precipitate with iron(III) solution. On standing, oxidation causes the iron (II) hydroxide to turn a brown-yellow colour due to gradual formation of iron(III) hydroxide.

   $Fe^{2+}(aq) + 2OH^-(aq) \rightarrow Fe(OH)_2(s)$

2. The thiocyanate ion gives a deep red colour with iron(III) but should give virtually no colour with iron(II). However, unless it is very pure and freshly prepared, iron(II) will probably give a faint red colour due to the presence of some iron(III).

3. Iron(III) oxidises iodide ions to iodine which gives the characteristic blue-black

THE ROYAL
SOCIETY OF
CHEMISTRY

colour with starch. Iron(II) should give no reaction unless it contains some iron(III).

$$2Fe^{3+}(aq) + 2I^-(aq) \rightarrow I_2(aq) + 2Fe^{2+}(aq)$$

4. The deep purple colour of manganate(VII) ions gradually diminishes as it is reduced by iron(II) whereas iron(III) has no effect.

$$MnO_4^-(aq) + 5Fe^{2+}(aq) + 8H^+(aq) \rightarrow Mn^{2+}(aq) + 5Fe^{3+}(aq) + 4H_2O(l)$$

5. The reaction of silver nitrate and iron(II) ions produces a glittering of metallic silver which is seen using a magnifying glass. There is no corresponding reaction with iron(III) ions.

Tips

These experiments can be done quickly so students might be encouraged to develop their explanations for the reactions. A book of data would be useful so that students can look up redox potentials. A biochemical development would be to consider the role of iron in haemoglobin and the types of iron compounds found in iron tablets (Iron(II) is required for haemoglobin, the +2 oxidation state being stabilised by complexation.)

The fact that many iron(II) compounds contain some iron(III) could form the basis of a discussion on the purities of chemical compounds.

Safety

Students must wear eye protection.

It is the responsibility of the teacher to carry out a risk assessment.

THE ROYAL
SOCIETY OF
CHEMISTRY

13. The transition elements

Topic

Transition elements – variable oxidation state, redox and precipitation reactions, and complex compounds. Chemistry and colour.

Level

Post-16.

Timing

20 min.

Apparatus (per group)

▼   One student worksheet

▼   One clear plastic sheet (*eg* ohp sheet)

▼   Magnifying glass.

Chemicals (per group)

Solutions contained in plastic pipettes, see p. 2

▼	Potassium chromate	$0.2 \ mol \ dm^{-3}$
▼	Potassium manganate(VII)	$0.2 \ mol \ dm^{-3}$
▼	Cobalt nitrate	$0.5 \ mol \ dm^{-3}$
▼	Ammonia solution	$3 \ mol \ dm^{-3}$
▼	Ammonium vanadate(V)	$0.2 \ mol \ dm^{-3}$
▼	Hydrochloric acid	$1 \ mol \ dm^{-3}$
▼	Sodium hydroxide	$1 \ mol \ dm^{-3}$
▼	Nickel nitrate	$0.5 \ mol \ dm^{-3}$
▼	Copper(II) sulphate	$0.2 \ mol \ dm^{-3}$
▼	Iron(II) sulphate	$0.2 \ mol \ dm^{-3}$
▼	Iron(III) nitrate	$0.2 \ mol \ dm^{-3}$
▼	Silver nitrate	$0.2 \ mol \ dm^{-3}$
▼	Potassium thiocyanate	$0.1 \ mol \ dm^{-3}$
▼	Potassium iodide	$0.2 \ mol \ dm^{-3}$
▼	Starch solution	(freshly made)
▼	Zinc metal granules.	

Observations

Vanadium

Bubbles (of hydrogen) are seen. The yellow colour of the ammonium vanadate (vanadium(V) ions) gradually changes (as the vanadium is reduced) to blue due to the formation of the vanadium(IV) ions (VO^{2+}). The colour changes to green due to vanadium(III) ions (V^{3+}) and possibly to lilac due to vanadium(II) ions (V^{2+}) (although this species is a strong reducing agent and is very air-sensitive).

Chromium

A red precipitate of silver chromate(VI) is seen. This is an interesting example of the modification of the colour of a coloured anion (yellow chromate(VI) by a colourless cation (silver(I)).

Manganese

The deep purple colour of the potassium manganate(VII) gradually fades, first to the brown manganese(IV) oxide then to the very pale pink manganese(II) ions. (Manganese(II) compounds in solution usually appear virtually colourless. However, a bottle of a solid manganese(II) salt – eg the sulphate – is pink.)

Iron

A yellowish colour (due to iodine) starts to form as the iron(II) oxidises the iodide. Addition of starch produces the characteristic intense blue-black colour of the starch–iodine complex.

Cobalt

The addition of one drop of ammonia gives a deep green precipitate. Addition of further ammonia gives a green or brown solution.

Nickel

The addition of sodium hydroxide produces a greenish precipitate of nickel(II) hydroxide.

Copper

The addition of ammonia gives a light blue precipitate of copper(II) hydroxide together with the deep blue tetra-amminocopper(II) ion.

Zinc

A white precipitate of zinc hydroxide is observed. (Zinc is not a transition metal because it only has one oxidation state in its compounds and the Zn^{2+} ion has a full d-sub-shell.)

Safety

Students must wear eye protection.
It is the responsibility of the teacher to carry out a risk assesment.

THE ROYAL
SOCIETY OF
CHEMISTRY

14. Reactions of transition elements

Topic

Transition elements – complex formation.

Level

Post-16.

Timing

20 min.

Apparatus (per group)

▼   One student worksheet

▼   One clear plastic sheet (*eg* ohp sheet).

Chemicals (per group)

Solutions contained in plastic pipettes, see p. 2

▼   Copper(II) sulphate 0.5 mol dm^{-3}

▼   Cobalt nitrate 0.5 mol dm^{-3}

▼   Nickel nitrate 0.5 mol dm^{-3}

▼   Magnesium nitrate 0.5 mol dm^{-3}

▼   Ammonia solution 3 mol dm^{-3}

▼   Sodium hydroxide 1 mol dm^{-3}.

Observations

The pink, green and blue colours of the cobalt, nickel and copper solutions contrast strongly with the colourless magnesium solution.

Reactions with ammonia

Cobalt

The addition of one drop of ammonia gives a deep green precipitate. Adding more ammonia gives a green or brown solution.

For cobalt, the Co^{2+} ion is much more stable than the strongly oxidising Co^{3+} ion. However, when Co^{2+} is reacted with ligands such as ammonia the hexa-amminocobalt(II) formed is oxidised to the stable hexa-amminocobalt(III) complex (this is the cause of any green-brown colour change). (Thus the presence of certain complexing ligands around Co^{3+} can greatly reduce its oxidising power.)

Nickel

The addition of one drop of ammonia slowly gives a light green precipitate – nickel(II) hydroxide. Adding more ammonia dissolves the precipitate to give a blue solution consisting of the hexa-amminonickel(II) complex.

Copper

The addition of one drop of ammonia with stirring gives a light blue precipitate of

copper(II) hydroxide. Adding more ammonia dissolves the precipitate to give the
deep blue tetra-amminocopper(II) ion.

Magnesium

The addition of ammonia gives a white precipitate of magnesium hydroxide. Adding
more ammonia produces no noticeable effect since the magnesium ion does not form
complexes with ammonia.

The reactions of the transition element ions with ammonia show that not only are
coloured products formed but that the colour and identity of the products depend on
the proportion of ammonia added.

Reactions with sodium hydroxide

Precipitates of metal hydroxides are seen in each case. The transition metal
hydroxides are coloured.

References

N. N. Greenwood and A. Earnshaw, *Chemistry of the elements.* Oxford: Pergamon,
1984.

Safety

Students must wear eye protection.

It is the responsibility of the teacher to carry out a risk assessment.

15. Chromium, molybdenum and tungsten

Topic

Transition elements – colours of ions, redox reactions, variable oxidation state.

Level

Post-16.

Timing

10 min.

Description

In this unusual experiment, some features of the chemistry of three transition elements are examined. The experiment illustrates aspects of colour, precipitate formation, changes in oxidation state and equilibria – all important concepts in transition metal chemistry. Most students will already be familiar with some of the chemistry of chromium through the oxidising properties of the dichromate ion, but not that of molybdenum and tungsten – the chemistry of these two elements is very complex.

Apparatus (per group)

▼　One student worksheet

▼　One clear plastic sheet (*eg* ohp sheet).

Chemicals (per group)

Solutions contained in plastic pipettes, see p. 2

▼　Potassium chromate　　　0.2 mol dm^{-3}

▼　Ammonium molybdate　　0.05 mol dm^{-3}

▼　Sodium tungstate　　　　0.2 mol dm^{-3}

▼　Hydrochloric acid　　　　1 mol dm^{-3}

▼　Sodium hydroxide　　　　1 mol dm^{-3}

▼　Iron(II) sulphate　　　　　0.2 mol dm^{-3}.

Discussion

Molybdenum and tungsten

These two transition metals, in the same group as chromium, are rarely encountered in experiments at secondary level although students should be familiar with the use of tungsten metal as filaments in light bulbs. (It has the highest melting point of any metal – 3422 ± 200 °C.) Molybdenum is an essential element in animal biochemistry and occurs in the enzyme xanthine oxidase which is involved in purine–adenine and guanine metabolism. Both metals are widely used in the steel industry where they are essential in the manufacture of high-speed steels for cutting tools.

　　Their solution chemistry is very complex – students might be surprised by the formula of ammonium molybdate!

　　The procedures described here illustrate changes in oxidation state (molybdenum)

THE ROYAL
SOCIETY OF
CHEMISTRY

and precipitation from acid solution (tungsten). The addition of a mild reducing agent (iron(II)) to a solution of molybdate produces a blue colour generally known as molybdenum blue. These are non-stoichiometric compounds, containing both oxides and hydroxides, the mean oxidation number of molybdenum being between 5 and 6. It has been suggested that Mo_3 atom clusters might be responsible for the blue colour.

The acidification of a solution of a tungstate produces a precipitate of tungsten(VI) oxide. Tungsten is therefore extracted from its ore (wolframite (Fe, Mn) WO_4) under alkaline conditions.

Observations

1. Students should observe that the solution of chromium (as chromate(III)) is coloured whereas those of molybdenum and tungsten are not. Discussion could ensue on whether this is significant, bearing in mind the general characteristic of transition elements that their compounds are coloured. Molybdenum and tungsten do form coloured compounds in other oxidation states and students should be aware of the blue copper(II) and white copper(I) salts.

2. Acidification of the yellow chromate(VI) solution produces an orange colour due to the formation of dichromate ions as the position of equilibrium is shifted:

$$2CrO_4^{2-} + 2H^+ \rightleftharpoons 2HCrO_4^- \rightleftharpoons Cr_2O_7^{2-} + H_2O$$

    yellow   intermediate   orange

    The addition of alkali removes the hydrogen ions and shifts the position of equilibrium back to the left and so the yellow colour is seen again.

3. Acidification of the molybdate(VI) solution produces a very interesting result. A white precipitate is formed at first. However, after a short while this precipitate starts to dissolve until eventually a colourless solution reforms. Molybdenum complexes are formed and, as these disproportionate, the species formed re-dissolve.

4. Acidification of the tungstate(VI) solution slowly produces a white precipitate of hydrated tungsten(VI) oxide, (WO_3).

    Whereas chromium and its compounds are soluble in acid solution, molybdenum and tungsten, and their compounds are precipitated at low pH but brought into solution at high pH. This is significant in the methods used to extract both molybdenum and tungsten from their ores.

5. The addition of iron(II) ions to chromate(VI) ions produces an orange colour. This is probably due to the effect of a change in pH. The iron(II) is oxidised to iron(III) and this, being slightly acidic, causes a shift in the position of equilibrium forming orange dichromate ions.

6. The addition of iron(II) ions – a mild reducing agent – to a molybdate solution produces a dark bluish colour sometimes known as molybdenum blue.

7. Addition of iron(II) ions to a tungstate solution produces a whitish precipitate of WO_3.

Reference

F. A. Cotton and G. Wilkinson, *Advanced inorganic chemistry*, 5th edn.
London: John Wiley, 1992.

THE ROYAL
SOCIETY OF
CHEMISTRY

Safety

Students must wear eye protection.

It is the responsibility of the teacher to carry out a risk assessment.

16. The reaction between hydrogen peroxide and dichromate ions

Topic

Transition elements – colours of ions, redox reactions, variable oxidation states.

Level

Post-16.

Timing

5 min.

Description

In this experiment dichromate(VI) ions are reduced to chromate(III) ions by hydrogen peroxide which is itself oxidised to oxygen gas. The experiment provides several points for student observation and illustrates an interesting redox reaction.

Apparatus (per group)

▼   One student worksheet

▼   One clear plastic sheet (*eg* ohp sheet).

Chemicals (per group)

Solutions contained in plastic pipettes, see p. 2

▼   Potassium dichromate   0.2 mol dm^{-3}

▼   Hydrogen peroxide   5% solution.

Observations

On adding the hydrogen peroxide solution, the reaction mixture immediately turns a deep blue colour. After a while bubbles are seen and the colour gradually fades to a pale blue-green due to hexa-aqua chromium(III) ions.
The reaction is:

$$Cr_2O_7^{2-}(aq) + H_2O_2(aq) + 8H^+(aq) \rightarrow 2Cr^{3+}(aq) + 5H_2O(l) + 2O_2(g)$$

Safety

Students must wear eye protection.
It is the responsibility of the teacher to carry out a risk assessment.

THE ROYAL
SOCIETY OF
CHEMISTRY

17. The chemistry of thiosulphate ions

Topic

Redox reactions; transition elements – redox reactions, catalysis, variable oxidation states.

Level

Post-16.

Timing

20 min.

Apparatus (per group)

▼   One student worksheet

▼   One clear plastic sheet (*eg* ohp sheet).

Chemicals (per group)

Solutions contained in plastic pipettes, see p. 2

▼	Sodium thiosulphate	0.1 mol dm^{-3}
▼	Silver nitrate	0.1 mol dm^{-3}
▼	Sodium chloride	0.1 mol dm^{-3}
▼	Potassium bromide	0.2 mol dm^{-3}
▼	Potassium iodide	0.2 mol dm^{-3}
▼	Iron(III) nitrate	0.1 mol dm^{-3}
▼	Copper(II) sulphate	0.2 mol dm^{-3}
▼	Iodine solution	0.05 mol dm^{-3} in 0.2 mol dm^{-3} KI.

Description

Sixth-form students usually encounter sodium thiosulphate in volumetric analysis where it is used as the titrant in a redox reaction with iodine:

$$I_2(s) + 2S_2O_3^{2-}(aq) \rightarrow S_4O_6^{2-}(aq) + 2I^-(aq)$$

However, in this experiment students investigate some of the many interesting reactions of sodium thiosulphate. These reactions illustrate many important chemical principles.

Observations

Part A
The brown colour of iodine is discharged as it is reduced by thiosulphate ions:

$$I_2(aq) + S_2O_3^{2-}(aq) \rightarrow 2I^-(aq) + S_4O_6^{2-}(aq)$$

Part B

The addition of halide ions to the silver nitrate solution produces precipitates of the silver halides – pale yellow (silver bromide) and deeper yellow (silver iodide). Silver bromide dissolves readily in sodium thiosulphate solution whereas silver iodide is less soluble. This could be used as a test to distinguish a bromide from an iodide.

$$Ag^+(aq) \quad + \quad X^-(aq) \quad \rightarrow \quad AgX(s)$$

Silver	Halide	Silver
ion	ion	halide

The dissolution of silver bromide in thiosulphate solution is used in the fixing stage in photographic developing. Here thiosulphate is used to dissolve unreacted silver bromide through the formation of soluble complexes such as $Ag(S_2O_3)_2{}^{3-}(aq)$.

Part C

The reaction of iron(III) with thiosulphate produces a deep violet complex anion, $Fe(S_2O_3)_2{}^-$. This decomposes slowly with the fading of the violet colour:

$$Fe(S_2O_3)_2{}^-(aq) \ + \ Fe^{3+}(aq) \rightarrow 2Fe^{2+}(aq) \ + \ S_4O_6{}^{2-}(aq)$$

The presence of copper(II) ions catalyses the decomposition reaction and the violet colour fades more rapidly.

Thiosulphate reduces Cu(II) to Cu(I) and complexes the Cu(I):

$$2S_2O_3{}^{2-} \ + \ 2Cu^{2+}(aq) \rightarrow 2Cu^+(aq) \ + \ S_4O_6{}^{2-}(aq)$$

$$2Cu^+(aq) \ + \ 2S_2O_3{}^{2-} \rightarrow Cu_2(S_2O_3)_2{}^{2-}(aq)$$

The characteristic blue colour of copper(II) fades leaving a colourless solution containing the complex ion $Cu_2(S_2O_3)_2{}^{2-}(aq)$.

Safety

Wear eye protection.
    It is the responsibility of the teacher to carry out a risk assessment.

THE ROYAL
SOCIETY OF
CHEMISTRY

18. The equilibrium of the cobalt chloride–water system

Topic

Equilibria, transition elements – ligands.

Level

Post-16.

Timing

20 min.

Description

In this experiment students observe colour changes in the cobalt chloride–water system by the addition of various reagents which affect the position of equilibrium.

Apparatus (per group)

▼ One plastic well-plate (24 wells) – *eg* Sigma ref: M9655.

Chemicals (per group)

Solutions contained in plastic pipettes, see p. 2

▼   Cobalt chloride (aqueous solution) $0.1 \ mol \ dm^{-3}$

▼   Cobalt chloride (ethanol solution) $0.1 \ mol \ dm^{-3}$

▼   Concentrated hydrochloric acid

▼   Concentrated sulphuric acid

▼   Potassium chloride $0.1 \ mol.dm^{-3}$

▼   Deionised water

▼   Potassium chloride powder.

Observations

The addition of water turns the blue ethanolic solutions of cobalt chloride from blue to pink. Adding potassium chloride solution to ethanolic cobalt chloride solution also causes a blue to pink colour change. Strongly dehydrating conditions (shifting the position of the equilibrium to the right) are required to generate the blue tetrachlorocobalt(II) complex from the pink hexa-aquacobalt(II) ion.

Reference

J. L. Mills and M. D. Hampton, *Microscale experiments for general chemistry*, 2nd edn. New York: McGraw-Hill, 1991.

Safety

Students must wear eye protection.
    It is the responsibility of the teacher to carry out a risk assessment.

THE ROYAL
SOCIETY OF
CHEMISTRY

19. Mass changes in chemical reactions

Topic

The nature of chemical reactions, scientific investigation.

Level

Pre-16.

Timing

10 min.

Description

In this experiment students measure the mass of various reactant solutions before and after reaction to see whether there has been any change in mass.

Apparatus (per group)

▼ One student worksheet

▼ Part of a well-plate (the three-well plate is cut from the standard 24-well plate using a hacksaw. A class set can be cut from a single well-plate.

▼ Access to a balance that reads to 0.01g.

Chemicals (per group)

Solutions contained in plastic pipettes, see p. 2

▼ Sodium carbonate   0.5 mol dm^{-3}

▼ Calcium nitrate   0.5 mol dm^{-3}

▼ Hydrochloric acid   1 mol dm^{-3}

▼ Magnesium ribbon

▼ Marble chips (small).

Observations

Students should find that there is a negligible difference in mass before and after mixing the sodium carbonate/calcium nitrate solution but there is some difference in mass in the magnesium or marble chip reaction with hydrochloric acid.

The success of this experiment depends on careful working by students and on the reliability of the balance and its proper use. Other combinations of substances could be examined and the experiment could be used as an investigation.

As an outcome of this experiment students should appreciate that matter is neither created nor destroyed in chemical reactions and that this is a very fundamental aspect of chemistry. It should also help them in balancing chemical equations!

Note

You will need a balance that reads to 0.01 g.

Safety

Students must wear eye protection.
It is the responsibility of the teacher to carry out a risk assessment.

THE ROYAL
SOCIETY OF
CHEMISTRY

20. Measuring density

Topic

Scientific methodology.

Level

Pre-16 and post-16.

Timing

20 min.

Description

In this experiment students determine and compare the density of tap water and sea-water using a small measuring cylinder and a sensitive balance.

Apparatus (per group)

▼ One student worksheet

▼ One 5 cm³ measuring cylinder

▼ One sheet of graph paper

▼ Access to a balance that reads to 0.01g.

Chemicals (per group)

Solutions contained in plastic pipettes, see p. 2

▼ Tap water

▼ Seawater.

Observations

Specimen results are given overleaf. From these results it is possible to distinguish the density of tap water and seawater using this method. The experiment is simple to do although students must work carefully to get good results.

A wealth of data are produced which when examined and interpreted show the application of mathematics to experimental science. In particular this shows the value of treating data graphically.

Students could be told that it was as a result of appreciating the significance of small differences in mass that important scientific discoveries have been made. For example, in 1894 William Ramsay and Lord Raleigh discovered the element argon in air by investigating the small but consistent discrepancy between the (higher) density of nitrogen obtained from air by removing the oxygen (atmospheric nitrogen) and nitrogen prepared from its compounds (*eg* heating ammonium nitrite solution).

Note

You will require a balance that reads to 0.01 g.

THE ROYAL
SOCIETY OF
CHEMISTRY

Specimen results measuring the densities of seawater and tap water

	Seawater			Tap water	
Vol (cm^3)	Mass (g)	Density (g cm^{-3})	Vol (cm^3)	Mass (g)	Density (g cm^{-3})
0.5	0.563	1.126	0.5	0.533	1.066
1.0	1.089	1.089	1.0	1.052	1.052
1.5	1.597	1.065	1.5	1.540	1.027
2.0	2.093	1.047	2.0	2.041	1.021
2.5	2.594	1.038	2.5	2.516	1.006
3.0	3.118	1.039	3.0	3.031	1.010
3.5	3.609	1.031	3.5	3.505	1.001
4.0	4.123	1.031	4.0	4.004	1.001
4.5	4.644	1.032	4.5	4.503	1.001
5.0	5.144	1.029	5.0	4.999	1.000
	mean density 1.053			mean density 1.018	

Extension

It may be interesting to extend this experiment to measure the relative densities of, for example, Coke® and Diet Coke®.

Safety

Students must wear eye protection.
It is the responsibility of the teacher to carry out a risk assessment.

THE ROYAL
SOCIETY OF
CHEMISTRY

21. Energy changes in neutralisation

Topic

Energy changes in reactions.

Level

Pre-16 and post-16.

Timing

20 min.

Description

In this experiment students do reactions at dropscale on thermometer strips and observe temperature changes.

Apparatus (per group)

▼   One student worksheet

▼   One thermometer strip.

Chemicals (per group)

Solutions contained in plastic pipettes, see p. 2

▼   Hydrochloric acid   2 mol dm^{-3}

▼   Sodium hydroxide   2 mol dm^{-3}

▼   Magnesium ribbon.

Observations

In both cases – the reaction of the piece of magnesium with the hydrochloric acid and the neutralisation reaction – it should be apparent that energy has been given out as heat since the numbers under the drops should illuminate. The highest green number can be regarded as the highest temperature reached.

It might be possible to obtain a value for the enthalpy change of the neutralisation of a strong acid by a strong base. Students know the concentration of the reagents and will be able to observe the temperature rise. They will need to know the volumes involved – the volume of one drop is *ca* 0.02 cm^3.

Safety

Students must wear eye protection.
It is the responsibility of the teacher to carry out a risk assessment.

THE ROYAL
SOCIETY OF
CHEMISTRY

22. Investigating temperature changes on evaporation of liquids

Topic

Scientific methodology.

Level

Pre-16 and post-16.

Timing

10 min.

Description

In this experiment students use the thermometer strip to examine temperature changes when drops of water, ethanol and ethoxyethane are placed on it.

Apparatus (per group)

▼   One student worksheet

▼   One temperature strip

▼   Pipettes.

Chemicals (per group)

▼   Ethoxyethane

▼   Deionised water

▼   Ethanol.

(**NB** Do not use propanone for this experiment – it attacks the temperature strip!)

Observations

Water, which forms well-defined droplets, produces very little, if any, change in temperature since the rate of evaporation is slow due to the high degree of hydrogen bonding.

With ethanol the drops spread out and a fall in temperature will be noted due to the higher rate of evaporation.

With ethoxyethane the drops evaporate very quickly and a marked drop in temperature is observed. This is consistent with the low boiling point and absence of hydrogen bonding between the molecules.

The energy changes accompanying changes of state are an important concept in science. One example is the addition of ice to cool drinks. Here it is the melting of ice that cools the drink rather than the contact of ice with the liquid.

Safety

Students must wear eye protection.

It is the responsibility of the teacher to carry out a risk assessment.

THE ROYAL
SOCIETY OF
CHEMISTRY

23. Investigating the effect of concentration on the rate of a chemical reaction

Topic

Reaction rates, scientific methodology.

Level

Pre-16 and post-16.

Timing

30 min.

Description

In this experiment students examine the effect of varying the concentration on the reaction between sodium thiosulphate and hydrochloric acid.

Apparatus (per group)

▼   One student worksheet

▼   One sheet of graph paper

▼   Stopclock

▼   Well-plate.

Chemicals (per group)

Solutions contained in plastic pipettes, see p. 2

▼   Sodium thiosulphate 0.1 mol dm^{-3}

▼   Hydrochloric acid 1 mol dm^{-3}

▼   Deionised water.

Observations

The time taken for the sulphur precipitate to appear is inversely proportional to the rate of the reaction and, for fixed concentrations of acid, students should find that the rate is proportional to $[S_2O_3^{2-}(aq)]$. The effect of varying the concentration of acid while keeping the concentration of thiosulphate constant can be used as a student investigation.

In each case students should measure the time until they can no longer see the black cross under the well-plate.

Safety

Students must wear eye protection.

It is the responsibility of the teacher to carry out a risk assessment.

Experiments to generate and test gases

The following experiments (24–30) involve generating and testing gases inside plastic petri dishes. The gases are: ammonia, carbon dioxide, chlorine, hydrogen sulphide, nitrogen dioxide and sulphur dioxide.

With the exception of carbon dioxide, these gases are toxic. Microscale offers the opportunity of doing experiments that would be difficult and hazardous to do at conventional scale without an efficient fume cupboard.

All the experiments use common pieces of apparatus – 9 cm plastic petri dishes (base and lid), a template, and plastic pipettes filled with the solutions required. In some tests (*eg* for chlorine) powdered solids are required.

In each case, the reagents to generate the gas are placed in a small plastic cup (called the reaction vessel in the student worksheet) cut from the end of a plastic pipette. This is then placed in the centre of the triangle in the petri dish overlying the template. The test solutions/solids are placed at the three corners of the triangle.

The general aim of these experiments is to generate sufficient gas to alter the appearance of the test solutions in a reasonably short time, without producing so much that it leaks out of the petri dish!

Apparatus (per group)

▼   Student information sheet and worksheets (one per gas)

▼   One 9 cm plastic petri dish (base + lid)

▼   One plastic pipette

▼   Scissors.

Chemicals (per group)

Solutions contained in plastic pipettes, see p. 2

Method

Students make the reaction vessel by cutting the end off the plastic pipette. If the gas being tested is chlorine, students can make a spatula for sampling the powders required in this experiment by cutting, at an angle of about 45°, another piece off the end of this pipette:

cut here

A data sheet on gases (Datagas) is provided for teachers which gives some background information on the detectability and toxicity of the gases.

THE ROYAL
SOCIETY OF
CHEMISTRY

Datagas – Exposure limits and odour thresholds for poisonous gases

Data obtained from Croner's *Substances hazardous to health*, Croner Publications: Kingston-upon-Thames, 1990)

(values in columns A, B and C are in parts per million)

	OES (8 hr) (A)	STEL (10 min) (B)	Approx. odour threshold (C)	B ÷ C
Chlorine	0.5	1	0.05	20
Sulphur dioxide	2	5	0.5	10
Nitrogen dioxide	3	5	5	1
Hydrogen sulphide	10	15	0.1	150
Ammonia	25	35	5	7
Carbon monoxide	50	300	–	–

OES – occupational exposure limit (8 hours)
STEL – short term exposure limit (10 minutes)
B ÷ C: STEL / odour threshold

The gases are listed in order of toxicity (the most toxic being chlorine). However, when the odour threshold is taken into account, the column B ÷ C can be considered to be a measure of how dangerous a gas is, the lower the value the more hazardous the gas. (I have not given data for carbon dioxide, which is non-toxic, and instead values are given for carbon monoxide, although it is not included in this set of experiments.)

Note on hydrogen sulphide

The approximate odour threshold for this gas is 0.1 ppm and the rotten eggs smell continues up to 20–30 ppm. As the concentration increases, however, the odour becomes sweetish and the gas diminishes the sense of smell until at about 150 ppm no odour is detectable.

THE ROYAL
SOCIETY OF
CHEMISTRY

24. Some reactions of ammonia

Topic

Gases.

Level

Pre-16 and post-16.

Timing

20 min.

Apparatus (per group)

▼ Student information sheet and worksheet

▼ One clear plastic sheet (*eg* ohp sheet)

▼ One 9 cm plastic petri dish (base + lid)

▼ One plastic pipette

▼ Scissors.

Chemicals (per group)

Solutions contained in plastic pipettes, see p. 2

▼ Ammonia solution $3 \ mol \ dm^{-3}$

▼ Full-range indicator solution diluted 1:1 with deionised water

▼ Copper(II) sulphate solution $0.2 \ mol \ dm^{-3}$

▼ Nessler's reagent (an alkaline solution of mercury iodide containing the complex ion HgI_4^-).

Method

Evaporation of ammonia gas from ammonia solution:

$$NH_3(aq) \rightarrow NH_3(g)$$

Tests

1. Full-range indicator solution turns blue-green.

2. Copper(II) sulphate solution turns hazy and then develops deep blue streaks as the tetra-amminocopper(II) ion is formed.

3. Nessler's reagent turns first yellow then brown. This is a very sensitive test for ammonia. The compound formed has the formula $(OHg_2NH_2)I$ and consists of covalent metal–non-metal bonds which might provide an interesting point for subsequent class discussion.

Safety

Students must wear eye protection. Nessler's reagent is toxic and contains mercury. It is the responsibility of the teacher to carry out a risk assessment.

THE ROYAL
SOCIETY OF
CHEMISTRY

25. Some reactions of carbon dioxide

Topic

Gases.

Level

Pre-16 and post-16.

Timing

20 min.

Apparatus (per group)

▼ Student information sheet and worksheet

▼ One clear plastic sheet (*eg* ohp sheet)

▼ One 9 cm plastic petri dish (base + lid)

▼ One plastic pipette

▼ Scissors.

Chemicals (per group)

Solutions contained in plastic pipettes, see p. 2

▼ Hydrochloric acid   1 mol dm^{-3}

▼ Barium nitrate solution   0.2 mol dm^{-3}

▼ Sodium hydroxide   0.5 mol dm^{-3}

▼ Small marble chips.

Method

The action of hydrochloric acid on marble chips generates carbon dioxide:

$$CaCO_3(s) + 2HCl(aq) \rightarrow CaCl_2(s) + CO_2(g) + H_2O(l)$$

Tests

1. The barium nitrate and sodium hydroxide drops should show no change.

2. The barium nitrate and sodium hydroxide mixture should turn cloudy owing to the formation of the very insoluble barium carbonate from the reaction of the (acidic) carbon dioxide gas with (alkaline) barium hydroxide.

Safety

Students must wear eye protection.
It is the responsibility of the teacher to carry out a risk assessment.

THE ROYAL
SOCIETY OF
CHEMISTRY

26. Some reactions of chlorine

Topic

Gases.

Level

Post-16.

Timing

20 min.

Apparatus (per group)

▼ Student information sheet and worksheet

▼ One clear plastic sheet (*eg* ohp sheet)

▼ One 9 cm plastic petri dish (base + lid)

▼ One plastic pipette

▼ Scissors.

Chemicals (per group)

Solutions contained in plastic pipettes, see p. 2

▼ Bleach

▼ Hydrochloric acid   1 mol dm^{-3}

▼ Sodium chloride   0.2 mol dm^{-3}

▼ Potassium bromide   0.2 mol dm^{-3}

▼ Potassium iodide   0.2 mol dm^{-3}.

Method

Bleach + hydrochloric acid generates chlorine:

$$HOCl(aq) + HCl(aq) \rightarrow Cl_2(g) + H_2O(l)$$

Results

The potassium iodide solution turns yellowish due to the liberation of iodine by the chlorine.
     The potassium bromide solution gradually turns pale yellow due to formation of bromine. No change occurs with the sodium chloride solution.

Safety

Students must wear eye protection.
     It is the responsibility of the teacher to carry out a risk assessment.

THE ROYAL
SOCIETY OF
CHEMISTRY

27. More reactions of chlorine

Topic

Gases.

Level

Post-16.

Timing

20 min.

Apparatus (per group)

▼   Student information sheet and worksheet

▼   One clear plastic sheet (*eg* ohp sheet)

▼   One 9 cm plastic petri dish (base + lid)

▼   One plastic pipette

▼   Scissors.

Chemicals (per group)

Solutions contained in plastic pipettes, see p. 2

▼   Bleach

▼   Hydrochloric acid $1 \ mol \ dm^{-3}$

▼   Zinc oxide powder

▼   Zinc sulphide powder

▼   Potassium iodide crystals.

Method

Bleach + hydrochloric acid generates chlorine:

$$HOCl(aq) + HCl(aq) \rightarrow Cl_2(g) + H_2O(l)$$

Results

The zinc sulphide takes on a yellowish tinge due to the liberation of elemental sulphur:

$$Cl_2(g) + ZnS(s) \rightarrow ZnCl_2(s) + S(s)$$

The potassium iodide becomes black due to the liberation of iodine. No change is observed with the zinc oxide – although it is probable that some reaction occurs to produce oxygen gas.

Safety

Students must wear eye protection.
   It is the responsibility of the teacher to carry out a risk assessment.

28. Some reactions of hydrogen sulphide

Topic

Gases.

Level

Pre-16 and post-16.

Timing

20 min.

Apparatus (per group)

▼ Student information sheet and worksheet

▼ One clear plastic sheet (*eg* ohp sheet)

▼ One 9 cm plastic petri dish (base + lid)

▼ One plastic pipette

▼ Scissors.

Chemicals (per group)

Solutions contained in plastic pipettes, see p. 2

▼ Hydrochloric acid	1 mol dm^{-3}
▼ Lead nitrate	0.5 mol dm^{-3}
▼ Potassium manganate(VII)	0.01 mol dm^{-3}
▼ Silver nitrate	0.2 mol dm^{-3}
▼ Sulphuric acid	1 mol dm^{-3}
▼ Zinc sulphide powder.	

Method

Zinc sulphide + hydrochloric acid generates hydrogen sulphide:

$$ZnS(s) + 2HCl(aq) \rightarrow ZnCl_2(s) + H_2S(g)$$

No bubbles of gas are seen and only small quantities of hydrogen sulphide are given off – sufficient, however, to carry out the tests.

Results

The lead nitrate solution gradually turns silvery due to the formation of lead sulphide by reaction of hydrogen sulphide and lead nitrate.

The silver nitrate solution reacts in a similar fashion although the reaction appears to be slower and the silver sulphide is less reflective than the lead sulphide.

The potassium manganate(VII) solution turns first brown and then, if sufficient gas is present, colourless, as it is reduced by the hydrogen sulphide:

$$2MnO_4^-(aq) + 5H_2S(g) + 6H^+(aq) \rightarrow 5S(s) + 2Mn^{2+}(aq) + 8H_2O(l)$$

THE ROYAL
SOCIETY OF
CHEMISTRY

Safety

Students must wear eye protection.

Hydrogen sulphide is an extremely poisonous gas but because it can be detected at very low concentrations it is much less dangerous than other gases such as carbon monoxide which, although less poisonous, cannot be detected by smell (see Datagas sheet p. 137).

It is the responsibility of the teacher to carry out a risk assessment.

29. Some reactions of nitrogen dioxide

Topic

Gases.

Level

Pre-16 and post-16.

Timing

20 min.

Apparatus (per group)

▼   Student information sheet and worksheet

▼   One clear plastic sheet (*eg* ohp sheet)

▼   One 9 cm plastic petri dish (base + lid)

▼   One plastic pipette

▼   Scissors.

Chemicals (per group)

Solutions contained in plastic pipettes, see p. 2

▼   Nitric acid (concentrated HNO_3)   diluted 1:1 with water *ca* 5M

▼   Full-range indicator solution   diluted 1:1 with deionised water

▼   Potassium iodide 0.2 mol dm^{-3}

▼   Potassium iodate(V) 0.1 mol dm^{-3}

▼   Potassium bromide 0.2 mol dm^{-3}

▼   Potassium bromate(V) 0.1 mol dm^{-3}

▼   Ammonia solution 3 mol dm^{-3}

▼   Copper turnings.

Method

Copper turnings + nitric acid generates first nitric oxide which then reacts with air to give nitrogen dioxide:

$$3Cu(s) + 8HNO_3(aq) \rightarrow 3Cu(NO_3)_2(aq) + 2NO(g) + 4H_2O(l)$$
$$\text{then: } 2NO(g) + O_2(g) \rightarrow 2NO_2(g)$$

Results

Full-range indicator turns from green to yellow-red indicating that nitrogen dioxide is an acidic gas.
    The iodate/iodide solution turns black due to:

$$IO_3^-(aq) + 5I^-(aq) + 6H^+(aq) \rightarrow 3I_2(g) + 3H_2O(l)$$

Also indicating the acidic nature of the gas.
A similar reaction occurs with bromide/bromate.

THE ROYAL
SOCIETY OF
CHEMISTRY

Safety

Students must wear eye protection.
It is the responsibility of the teacher to carry out a risk assessment.

THE ROYAL
SOCIETY OF
CHEMISTRY

30. Some reactions of sulphur dioxide

Topic

Gases.

Level

Pre-16 and post-16.

Timing

20 min.

Apparatus (per group)

▼   Student information sheet and worksheet

▼   One clear plastic sheet (*eg* ohp sheet)

▼   One 9 cm plastic petri dish (base + lid)

▼   One plastic pipette

▼   Scissors.

Chemicals (per group)

Solutions contained in plastic pipettes, see p. 2

▼   Hydrochloric acid   1 mol m^{-3}

▼   Potassium iodide   0.2 mol dm^{-3}

▼   Potassium iodate(V)   0.1 mol dm^{-3}

▼   Potassium manganate(VII)   0.01 mol dm^{-3}

▼   Full-range indicator solution   diluted 1:1 with deionised water

▼   Sulphuric acid   1 mol dm^{-3}

▼   Sodium sulphite powder.

Method

Sodium sulphite + hydrochloric acid generates sulphur dioxide:

$$Na_2SO_3(s) + 2HCl(aq) \rightarrow 2NaCl(s) + SO_2(g) + H_2O(l)$$

Results

The iodide/iodate mixture turns black due to liberation of iodine:

$$IO_3^-(aq) + 5I^-(aq) + 6H^+(aq) \rightarrow 3I_2(g) + 3H_2O(l)$$

If sufficient sulphur dioxide is produced and the solution contains excess acid, the potassium manganate(VII) solution is decolorised:

$$8H^+(aq) + 5e^- + MnO_4^-(aq) \rightarrow Mn^{2+}(aq) + 4H_2O(l)$$

THE ROYAL
SOCIETY OF
CHEMISTRY

However, with less sulphur dioxide and therefore less acid, the brown manganese(IV) oxide is formed:

$$4H^+(aq) + MnO_4^-(aq) + 3e^- \rightarrow MnO_2(s) + 2H_2O(l)$$

Full-range indicator turns from green to yellow indicating that sulphur dioxide is an acidic gas.

Safety

Students must wear eye protection.
It is the responsibility of the teacher to carry out a risk assessment.

31. Oxygen and methylene blue

Topic

Organic chemistry, redox reactions, dyes and colour chemistry.

Level

Pre-16 and post-16.

Timing

15 min.

Description

In this experiment students generate oxygen gas by the reaction between hydrogen peroxide and potassium manganate(VII), and then test for the gas by bubbling it into a solution of the reduced form of methylene blue dye, turning the solution blue.

Apparatus (per group)

▼   One student worksheet

▼   One 10 cm³ beaker

▼   One plastic pipette (standard form)

▼   One piece of rubber tubing, *ca* 10 cm long

▼   Scissors.

Chemicals (per group)

Solutions contained in plastic pipettes, see p. 2

▼   Hydrogen peroxide   5% solution

▼   Potassium manganate(VII)   0.1 mol dm⁻³

▼   Methylene blue solution (colourless, leuco form of dye)

▼   Glucose.

Dissolve 4 g of potassium hydroxide pellets in 150 cm³ of deionised water in a plastic bottle or stoppered 250 cm³ conical flask. Allow to cool and add 5 g of glucose powder. Add 3–4 drops of methylene blue solution (0.25 g in 1000 cm³ of deionised water or Aldrich cat. no. 31,911-2). The blue solution should become colourless on standing a few minutes but will turn blue when shaken.

Observations

This experiment is a little tricky to perform and students will need to practice it first! The hydrogen peroxide and potassium manganate(VII) react together vigorously to produce oxygen gas. The colourless solution of methylene blue should turn blue quickly when the oxygen gas is directed into it.

Students are given the structures of the oxidised and reduced forms of methylene blue and are asked to say which is which. The oxidised (blue) form contains conjugated double and single bonds throughout the whole molecule whereas in the colourless form the delocalised electron systems are isolated from each other. The structures are given overleaf.

THE ROYAL
SOCIETY OF
CHEMISTRY

Reference

D. Barton and W. D. Ollis, *Comprehensive organic chemistry*, vol 4, pp1102–1107.
Oxford: Pergamon, 1979.
This book gives an interesting account of the dibenzo-1,4-thiazines, of which
methylene blue is a member.

Safety

Students must wear eye protection.
It is the responsibiliy of the teacher to carry out a risk assessment.

THE ROYAL
SOCIETY OF
CHEMISTRY

32. A microscale study of gaseous diffusion

Topic

Diffusion as evidence for particles, diffusion equations, gases, transition elements (redox reactions, catalysis and variable oxidation states).

Level

Pre-16 and post-16.

Timing

30–40 min.

Description

This experiment looks at the spread of ammonia and chlorine gases as a result of their interaction with copper(II) sulphate and potassium iodide/starch solutions. The experiment is done using drops of solutions placed on a clear plastic sheet with the top of a well-plate as a lid. The experiment is suitable for various levels with differing degrees of interpretation.

Apparatus (per group)

▼ One student worksheet

▼ One clear plastic sheet (*eg* ohp sheet)

▼ Two well-plate lids (24 well size, *eg* Sigma ref: M 9655)

▼ Two plastic pipettes

▼ Magnifying glass.

Chemicals (per group)

Solutions contained in plastic pipettes, see p. 2

▼ Concentrated ammonia solution

▼ Potassium iodide   0.2 mol dm^{-3}

▼ Hydrochloric acid   1 mol dm^{-3}

▼ Copper(II) sulphate   0.5 mol dm^{-3}

▼ Starch solution

▼ Bleach.

Observations

In the ammonia experiment the drops of copper(II) sulphate solution first turn opaque blue due to the formation of copper(II) hydroxide and then gradually develop dark blue blotches and streaks as the tetra-amminocopper(II) ion is formed. The colour changes and patterns are seen using a magnifying glass. The fact that patterns are seen at all and that the colour changes are not uniform suggests movement within the drops.

THE ROYAL
SOCIETY OF
CHEMISTRY

In the chlorine experiment the liberated iodine gives a blue-black colour with the starch. In both experiments there is a gradation of colour change which depends on the rate of diffusion of the gas. It might be possible to compare the rate of diffusion of these two gases. Ammonia with a relative molecular mass of 17 should be faster than chlorine with a relative molecular mass of 71. The rates of evaporation of ammonia from concentrated ammonia solution and the rate of the generation of chlorine from the reaction between bleach and hydrochloric acid may be different. Moreover, the sensitivities of the respective drops to show colour changes on reaction with the gases may also be different.

Students may also be encouraged to develop their understanding of the properties of gases from this experiment. For example, the large difference in relative molecular mass of ammonia and chlorine manifests itself in differences in rates of diffusion by combining Avogadro's Hypothesis (equal volumes of all gases contain, under equal conditions, equal numbers of molecules) and Graham's Law of Gaseous Diffusion (the rate of diffusion of a gas is inversely proportional to the square root of its density).

In the chlorine experiment, the liberated iodine tends to stain the plastic sheet if left in contact with it for a while. Since the amount of iodine liberated depends on the amount of chlorine present a note of the rate of chlorine diffusion is recorded on the plastic sheet. This could be shown on an overhead projector or displayed on a wall.

Reference

J.Chem.Ed., 1989, **66**, 436.

Safety

Students must wear eye protection.

It is the responsibility of the teacher to carry out a risk assessment.

THE ROYAL
SOCIETY OF
CHEMISTRY

33. Acid – base neutralisation

Topic

Scientific methodology, acids/bases/neutralisation/chemical techniques.

Level

Pre-16 and post-16.

Timing

30 min.

Description

This experiment involves acid/base neutralisation using microscale titration apparatus (see p. 7)

Students do the titration by filling the 'burette' with hydrochloric acid and placing 1 cm³ of sodium hydroxide solution in a 10 cm³ beaker. The aim is to calculate the exact concentration of the sodium hydroxide solution. It is important that students do not use more than 1 or 2 drops of indicator otherwise the accuracy of the result may be affected (since indicators, being weak acids or bases themselves, will consume small amounts of acid or base).

Apparatus (per group)

▼ Microscale titration apparatus, see p. 7

▼ One 1 cm³ pipette + pipette filler

▼ Two 10 cm³ beakers.

Chemicals (per group)

▼	Hydrochloric acid	0.10 mol dm⁻³
▼	Sodium hydroxide solution	*ca* 0.1 mol dm⁻³
▼	Phenolphthalein indicator solution.	

Safety

Students must wear eye protection.

It is the responsibility of the teacher to carry out a risk assessment.

34. Measuring an equilibrium constant

Topic

Scientific methodology, reversible reactions.

Level

Post-16.

Timing

30 min.

Description

This experiment is a microscale version of experiment 11.2 in the Nuffield A-Level Chemistry course (3rd edn. 1994). It uses the microscale titration apparatus (see p. 7) to measure the equilibrium constant of the redox reaction between silver(I) and iron(II):

$$Ag^+(aq) + Fe^{2+}(aq) \rightleftharpoons Ag(s) + Fe^{3+}(aq)$$

One of the main advantages of this microscale version is that it uses less silver solution than the existing method and is therefore more economical.

Apparatus (per group)

▼ Microscale titration apparatus, see p. 7

▼ One stoppered flask (eg 5–10 cm³)

▼ Two 2 cm³ pipettes

▼ One 10 cm³ beaker.

Chemicals (per group)

▼ Iron(II) sulphate solution   0.10 mol dm⁻³

Dissolve 2.780 g of Analar grade $FeSO_4.7H_2O$ in 50 cm³ of 0.5 M sulphuric acid and make up to 100 cm³ with deionised water in a volumetric flask.

▼ Silver nitrate solution   0.10 mol dm⁻³

Dissolve 0.849 g of $AgNO_3$ in deionised water and make up to 50 cm³ in a volumetric flask. Store in the dark if this solution is not to be used immediately (it will be useable for several days). This is sufficient solution for 20–25 pairs of students to do the experiment. The amounts may be scaled down if appropriate.

▼ Potassium thiocyanate solution   0.020 mol dm⁻³

Dissolve 0.194 g of potassium thiocyanate in deionised water and make up to 100 cm³ in a volumetric flask.

Observations

A white precipitate of silver thiocyanate forms until the end-point when a permanent red colour due to $Fe(SCN)^{2+}$ is seen.

THE ROYAL
SOCIETY OF
CHEMISTRY

Results

The Nuffield teachers' guide does not give a value for K_c and merely says that the results are rather variable, due in part to the difficulty of maintaining a pure solution of iron(II) sulphate.

The Nuffield textbook states that the silver nitrate/iron(II) sulphate mixture should be allowed to stand overnight for equilibrium to be established. These conditions have been adhered to for this microscale titration although it is possible that less time is required (this could form the basis of an interesting student project, *ie* establishing the time taken to reach equilibrium). Nevertheless, students should set up the stoppered flask at the end of one lesson and do the titration at the next.

Reference

Nuffield advanced science – chemistry, students' book and teachers' guide. 3rd edn. Harlow: Longmans, 1994.

Safety

Students must wear eye protection.

It is the responsibility of the teacher to carry out a risk assessment.

THE ROYAL
SOCIETY OF
CHEMISTRY

35. Finding out how much salt there is in seawater

Topic

Earth science, scientific methodology and chemical analysis.

Level

Pre-16 and post-16.

Timing

30 min.

Description

In this experiment students determine the chloride content of seawater by using a microscale version of the Mohr titration – titrating standard silver nitrate against the seawater using potassium chromate as indicator. Having found the chloride content the assumption is then made that all the chloride is associated with sodium ions and hence the percentage by mass of sodium chloride in the water is calculated.

A particular advantage of this microscale method is the reduction in cost of the experiment since far smaller quantities of silver nitrate solutions are used compared with conventional titrations.

Apparatus (per group)

▼   Microscale titration apparatus (see p. 7)

▼   Two 10 cm³ beakers

▼   One 1cm³ pipette + pipette filler.

Chemicals (per group)

▼   Plastic pipette filled with potassium chromate indicator solution

▼   Silver nitrate solution

▼   Sample(s) of seawater (or artificial seawater made by dissolving *ca* 3 g sodium chloride in 100 cm³ water).

Silver nitrate solution

Dissolve *ca* 4 g of silver nitrate, accurately weighed, in deionised water and make up to 50 cm³ in a volumetric flask. This solution will be *ca* 0.5 mol dm⁻³ and should be sufficient for 20–30 titrations. If fewer titrations are envisaged the quantities can be scaled down to reduce cost – *eg* by dissolving 2 g of silver nitrate in 25 cm³ of water.

Potassium chromate solution

Dissolve 2 g of K_2CrO_4 in 50 cm³ of deionised water. This solution will be *ca* 0.2 mol dm⁻³.

Observations

Adding one drop of potassium chromate indicator gives the seawater a yellowish colour prior to the titration. During the titration a white precipitate of silver chloride

THE ROYAL
SOCIETY OF
CHEMISTRY

forms until the end-point when a permanent red colour of silver chromate appears –
silver chloride is less soluble than silver chromate.

A titre of *ca* 1.5 cm³ should be obtained.

Note

Teachers may wish to use this experiment without the section on results and
calculations.

Students could be asked how they could show that the salt in seawater was
sodium chloride and not, say, potassium chloride (flame test) and whether there may
be anything else in the seawater that could have interfered with their titration.

References

J. Chem.Ed., 1992, **69,** 830.

Safety

Students must wear eye protection.

It is the responsibility of the teacher to carry out a risk assessment.

THE ROYAL
SOCIETY OF
CHEMISTRY

36. Measuring the amount of vitamin C in fruit drinks

Topic

Food, scientific methodology. Quantitative chemistry/mole calculations.

Level

Pre-16 and post-16.

Timing

20 min.

Description

In this experiment students use the microscale titration technique to measure the amount of vitamin C (ascorbic acid) in fruit drinks. The basis of the measurement is as follows.

A known excess amount of iodine is generated by the reaction between iodate, iodide and sulphuric acid:

$$IO_3^-(aq) + 5I^-(aq) + 6H^+(aq) \rightarrow 3I_2(aq) + 3H_2O(aq)$$

A measured amount of fruit drink is added. The ascorbic acid in the drink reacts quantitatively with some of the iodine:

Ascorbic acid		Dehydroascorbic acid

The excess iodine is then titrated against standard thiosulphate solution:

$$I_2 + 2S_2O_3^{2-} \rightarrow S_4O_6^{2-} + 2I^-$$

Chemicals (per group)

▼   Sodium thiosulphate

▼   Potassium iodate

▼   Potassium iodide

Solutions contained in plastic pipettes, see p. 2

▼   Starch solution (freshly made)

▼   Sulphuric acid 1 mol dm^3

▼   Sample(s) of fruit juice.

Apparatus (per group)

▼ One student worksheet

▼ Microscale titration apparatus (see p. 7)

▼ One 1 cm³ pipette (glass)

▼ One 2 cm³ pipette (glass)

▼ Pipette filler

▼ One 25 cm³ beaker

▼ One 5 cm³ measuring cylinder

▼ One 10 cm³ beaker (for filling titration apparatus).

Stock solutions

1. Sodium thiosulphate solution   0.010 mol dm⁻³

    Weigh out, accurately, *ca* 0.620 g of $Na_2S_2O_3.5H_2O$, dissolve in deionised water and make up to 250 cm³ in a volumetric flask. Store this stock solution in a dark glass bottle.

2. Potassium iodate solution   0.001 mol dm⁻³

    Weigh out, accurately, *ca* 0.054 g of KIO_3, dissolve in deionised water and make up to 250 cm³ in a volumetric flask.

3. Potassium iodide solution   0.005 mol dm⁻³

    Weigh out 0.21 g of KI, dissolve in deionised water and make up to 250 cm³ with deionised water.

Note

The reaction to generate the iodine is based on using an accurately known volume of the potassium iodate solution (the concentration of which is accurately known). The potassium iodide solution and the sulphuric acid are added in slight excess and thus the concentrations of these solutions is not critical.

Observations

The titre volume should be in the range 0.5–1 cm³, the disappearance of the blue-black colour marking the end-point.

This experiment offers possibilities for assessing students' abilities in following instructions and/or processing results.

A survey of a range of fruit drinks (and maybe other products containing vitamin C) could form the basis of a class project or as an activity for a school or college chemistry club.

Specimen result and calculation

Volume of thiosulphate delivered during the titration = 0.74 cm^3.

Concentration of thiosulphate = 0.010 mol dm^{-3}.

Therefore number of moles thiosulphate =

$$\frac{0.74 \times 0.01}{1000} = 7.4 \times 10^{-6}$$

Therefore the number of moles of iodine that this reacted with during the titration = 3.7 x 10^{-6}.

The total number of moles of iodine produced in the reaction between iodate, iodine and sulphuric acid based on using 2 cm^3 of iodate with a concentration of 0.0012 mol dm^{-3} =

$$\frac{3 \times 2 \times 0.0012}{1000} = 7.2 \times 10^{-6}$$

Therefore the number of moles of iodine which reacted with the ascorbic acid =

7.2 x 10^{-6} – 3.7 x 10^{-6} = 3.5 x 10^{-6}

Since 1 mole of iodine reacts with 1 mole of ascorbic acid then the number of moles of ascorbic acid is also 3.5 x 10^{-6}.

The volume of the fruit juice used was 1 cm^3. Therefore the number of moles of ascorbic acid in 1000 cm^3 = 3.5 x 10^{-3}.

The relative molar mass of ascorbic acid = 174.12 g. Therefore mass of ascorbic acid (in 1000 cm^3) = 174.12 g x 3.5 x 10^{-3} = 0.609 g.

The vitamin C content of the fruit drink = 61 mg per 100 cm^3.

Reference

J.Chem.Ed., 1992, **69**, A213-4.

Note

Instead of generating the iodine *in situ*, it is possible to use standard iodine solution in this procedure. This would need to be diluted to give an aliquot containing 7.2 x 10^{-6} moles of iodine (see above) for each determination.

Safety

Students must wear eye protection.
    It is the responsibility of the teacher to carry out a risk assessment.

THE ROYAL
SOCIETY OF
CHEMISTRY

37. Using a microscale conductivity meter

Topic

Solutions – conductivity, ions. Metals – conducting electricity.

Level

Pre-16 and post-16.

Timing

15 min.

Description

In this experiment students use the conductivity meter to test the conductivity of solids (*eg* metals) or solutions. The test is very easy to do and virtually any material can be examined. Students will need to be careful about cross-contamination when testing solutions.

Apparatus (per group)

▼   One clear plastic sheet (*eg* ohp sheet)

▼   Conductivity meter (see p. 10).

Chemicals (per group)

▼   Copper(II) sulphate solution

▼   Sodium chloride solution

▼   Tap water

▼   Deionised water

▼   Sugar solution

▼   Copper foil

▼   Aluminium foil

▼   Iron nail

▼   Pencil lead.

Observations

Metals and solutions/liquids that contain ions should cause the light emitting diode (LED) to shine. This experiment provides a quick and simple method for testing conductivity. The LED will light for any substance – whether liquid or solid – that conducts.

Safety

Students must wear eye protection.
    It is the responsibility of the teacher to carry out a risk assessment.

THE ROYAL
SOCIETY OF
CHEMISTRY

38. Electrolysis using a microscale Hoffman apparatus

Topic

Electrolysis.

Level

Pre-16 and Post-16.

Timing

25 min.

Description

In this experiment students use a microscale Hoffman apparatus to investigate the electrolysis of sodium sulphate solution.

Apparatus (per group)

▼   Clamp and stand

▼   Microscale Hoffman apparatus (see p. 13)

▼   One 9 volt battery and leads with crocodile clips

▼   Plastic pipette

▼   Blu-Tack®

▼   One 100 cm³ beaker.

Chemicals (per group)

▼   Sodium sulphate solution 0.2 mol dm^{-3}

▼   Bromothymol blue indicator.

Observations

Streams of bubbles are seen at each electrode. The colour of the solution around the cathode gradually turns blue due to the formation of sodium hydroxide. The solution around the anode becomes greenish-yellow. If the tops of the pipettes are sealed with Blu-Tack® the volume of gas collecting above the cathode (hydrogen) is seen to be greater than that collecting above the anode (oxygen). If left connected for long enough the ratio of the volumes corresponds to the 2:1 ratio of hydrogen:oxygen in water. The shortened pipette (see p. 2) can be used to sample the hydrogen gas and to test it by blowing it into a flame (it 'pops').

Safety

Students must wear eye protection.
It is the responsibility of the teacher to carry out a risk assessment.

39. The determination of copper in brass

Topic

Metals–chemical analysis.

Level

Post-16.

Timing

25 min.

Description

In this experiment students dissolve some brass in nitric acid and compare the colour of the solution against standard copper solutions in a well-plate. This experiment has possibilities for use as an assessed practical. Two versions of the student worksheet are given (versions A and B).

In version A students are guided through the calculations at the end. This version could be used to assess skills in doing the experiment/following instructions. In version B no help is given with the calculations. This version could be used to assess skills in treatment of results.

Apparatus (per group)

▼ One student worksheet and one sheet of white paper

▼ Access to a balance

▼ Access to a fume cupboard

▼ One 10 cm³ beaker

▼ One 10 cm³ volumetric flask

▼ One plastic well-plate (24 well) – *eg* Sigma ref: M9655

▼ One plastic pipette – *eg* Aldrich ref: Z13,503-8, fine-tip.

Chemicals (per group)

Solutions contained in plastic pipettes, see p. 2

▼ Nitric acid $5 \ mol \ dm^{-3}$

▼ Deionised water

▼ Copper nitrate solution $0.50 \ mol \ dm^{-3}$

▼ Brass turnings

Observations

The brass dissolves quickly to form a blue solution. This colour is due to the copper present in the brass.

(This part of the experiment must be done in a fume cupboard since nitrogen dioxide is formed.) The intensity of the colour of this solution should lie within the range of intensities of colour of the standard solutions. Students find the nearest colour match and then calculate the copper content of the brass.

THE ROYAL
SOCIETY OF
CHEMISTRY

Discussion

Most brass contains about 60% copper (the remainder being zinc). Brass forms an interesting subject for a discussion on the structure of metals and alloys.

Copper metal has a face-centered cubic structure (fcc) while the structure of zinc is hexagonal. As zinc is added to copper it substitutes in the lattice to form a distorted fcc structure (zinc atoms are *ca* 13% larger than copper). This distorted structure is difficult to deform and accounts for the greater strength of brass compared to pure copper.

When the zinc content reaches about 36% a new body centered cubic phase appears and the strength increases markedly although the ductility is reduced. The optimum properties of strength and ductility for most uses of brass occur at about 40% zinc.

Reference

A. Street and W. Alexander, *Metals in the service of man*, 10th edn. London: Penguin, 1994.

Safety

Students must wear eye protection.

It is the responsibility of the teacher to carry out a risk assessment.

THE ROYAL
SOCIETY OF
CHEMISTRY

40. Observing the lowering of a melting point

Topic

Melting point.

Level

Post-16.

Timing

15 min.

Description

Pure compounds have characteristic sharp melting points. The presence of impurities lowers the melting point and broadens its range. In this experiment students mix two solid organic compounds with low melting points and observe the mixture melting.

OH

Hydroxybenzene
(phenol)

CH₃

OH

H₃C   CH₃

Menthol

Apparatus (per group)

▼   One plastic pipette (standard form, *eg* Aldrich ref: Z13,500-3)

▼   One fine-tipped plastic pipette (standard form, *eg* Aldrich ref: Z13,503-8)

▼   One plastic pipette (for mixing the solids in the dish)

▼   One plastic petri dish (*eg* 5.5 cm size from Aldrich).

Chemicals (per group)

▼   Hydroxybenzene (phenol)

▼   Menthol

▼   Sodium hydroxide solution.

Tip

Students should use the method for sampling a bottle of hydroxybenzene (phenol) (see p. 5).

Discussion

This experiment illustrates not only the lowering of melting point, but also several other important points of relevance to post-16 organic chemistry courses.

Safety

It is possible to do experiments at microscale with hazardous organic compounds.

Structure

The phenolic hydroxyl group can be contrasted with the alcoholic hydroxyl group on the menthol.

Stereoisomerism

Menthol consists of enantiomers.

Intermolecular bonding

Nuclear magnetic resonance and infrared evidence of shifts in O–H absorptions suggest hydrogen bonding between hydroxybenzene and menthol molecules.

During trials of this experiment, one teacher used camphor (mp 176 °C) instead of menthol. Liquifaction occurred readily, which, considering the high melting point of camphor, was very interesting. A wide range of organic solids could be examined using this procedure which could form the basis of an interesting student project.

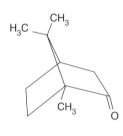

Camphor

Reference

J.Chem. Ed., 1990, **67**, 156.

Safety

Students must wear eye protection. Hydroxybenzene (phenol) is hazardous and gloves should be worn.

It is the responsibility of the teacher to carry out a risk assessment.

THE ROYAL
SOCIETY OF
CHEMISTRY

41. Properties of stereoisomers

Topic

Stereochemistry.

Level

Post-16.

Timing

5 min.

Description

In this experiment students detect the differences in smell of each enantiomer absorbed on cotton wool inside small sample bottles. To prepare these place a small quantity of cotton wool into each bottle and then add 10 drops of the stereoisomer. The bottles can then be passed around the classroom.

(R) - (+) - Limonene   (S) - (-) - Limonene

Apparatus (per group)

▼   Two plastic bottles

▼   Cotton wool.

Chemicals (per group)

▼   (R)-(+)-Limonene

▼   (S)-(–)-Limonene.

Extension

Students could obtain small quantities of the stereoisomers of limonene in natural fruits by carrying out steam distillation of the peel of citrus fruits such as oranges and lemons and comparing the odours against the standards.

Safety

Students must wear eye protection.
   It is the responsibility of the teacher to carry out a risk assessment.

THE ROYAL
SOCIETY OF
CHEMISTRY

42. Properties of the carvones

Topic

Stereochemistry.

Level

Post-16.

Timing

5 min.

Description

In this experiment students detect the differences in smell of each enantiomer absorbed on cotton wool inside a small sample bottles. To prepare these, place a small amount of cotton wool into each bottle and then add ten drops of the stereoisomer. The bottles can then be passed around the classroom.

(*R*) - (-) - Carvone   (*S*) - (+) - Carvone

Chemicals (per group)

▼   (*R*)-(-)-Carvone

▼   (*S*)-(+)-Carvone.

Extension

Students could obtain small quantities of chewing gum – which contains spearmint – and caraway seeds and compare the smells of these.

Note

Limonene, a terpene occurring in orange and lemon oils, and carvone are structurally very similar. It can be pointed out to students that with the limonene steroisomers the *R*-enantiomer is dextrorotatory and the *S*-enantiomer is laevorotatory. With the carvones the opposite is true.

Safety

Students must wear eye protection.
  It is the responsibility of the teacher to carry out a risk assessment.

THE ROYAL
SOCIETY OF
CHEMISTRY

43. The formation of 2,4,6-trichlorohydroxybenzene by the reaction between hydroxybenzene and chlorine gas

Topic

Reactions of organic molecules, reactions using chlorine.

Level

Post-16.

Timing

20 min.

Description

This experiment is done in a plastic petri dish. 2,4,6-Trichlorohydroxybenzene (2,4,6-trichlorophenol, TCP) is detected by its distinctive antiseptic smell.

Apparatus (per group)

▼   One clear plastic sheet (eg ohp sheet)

▼   One plastic petri dish (*eg* 5.5 cm diameter)

▼   One plastic pipette.

Chemicals (per group)

Solutions contained in plastic pipettes, see p. 7

▼   Bleach

▼   Hydrochloric acid   1 mol dm^{-3}

▼   Sodium hydroxide   1 mol dm^{-3}

▼   Hydroxybenzene (phenol).

Tip

Use method of sampling a bottle of hydroxybenzene (phenol) (p. 5).

Method

Students cut off the end of the plastic pipette to make a reaction vessel for generating the chlorine gas using bleach and hydrochloric acid. After about 15 min the lid is taken off the petri dish and a strong antiseptic smell should be detected. At the end of the experiment students add drops of sodium hydroxide solution to the hydroxybenzene (phenol)/TCP residue to dissolve it and then mop up the contents of the petri dish with a tissue.

Safety

Students must wear eye protection. Hydroxybenzene (phenol) is hazardous and gloves should be worn.

It is the responsibility of the teacher to carry out a risk assessment.

THE ROYAL
SOCIETY OF
CHEMISTRY

44. The chemical properties of hydroxybenzene

Topic

Organic compounds, chemical properties.

Level

Post-16.

Timing

20 min.

Description

In this experiment students observe and interpret some of the chemical reactions of hydroxybenzene (phenol).

Apparatus (per group)

▼ One clear plastic sheet (*eg* ohp sheet)

▼ One plastic petri dish (9 cm diamater).

Chemiclas (per group)

Solutions contained in plastic pipettes, see p. 2

▼ Deionised water

▼ Full-range indicator solution – diluted 1:1 with deionised water

▼ Nitric acid   1 mol dm^{-3}

▼ Iron(III) nitrate   0.1 mol dm^{-3}

▼ Sodium carbonate   1 mol dm^{-3}

▼ Sodium hydroxide   1 mol dm^{-3}

▼ Hydrochloric acid   1 mol dm^{-3}

▼ Hydroxybenzene (phenol).

The students should sample the bottle of hydroxybenzene (phenol) using the method on p. 5.

Observations

1. Hydroxybenzene (phenol) is partially soluble in water and oily drops should be observed. An acidic solution is formed and the indicator solution should turn red.

2. The mixture turns dark as the hydroxybenzene (phenol) reacts with the nitric acid to give a mixture of nitrophenols.

3. A violet coloration is seen which is characteristic of the reaction between iron(III) and phenolic OH groups.

4.   No reaction is observed with sodium carbonate solution indicating that hydroxybenzene (phenol), although acidic, is not a strong enough acid to liberate carbon dioxide from carbonates.

5.   Hydroxybenzene (phenol) dissolves readily in sodium hydroxide to form sodium phenoxide. The hydroxybenzene (phenol) is liberated and oily drops should be observed when this solution is acidified with hydrochloric acid.

Safety

Students must wear eye protection. Hydroxybenzene (phenol) is hazardous and gloves should be worn.

It is the responsibility of the teacher to carry out a risk assessment.

THE ROYAL
SOCIETY OF
CHEMISTRY

45. A test to distinguish between methanol and ethanol

Topic

Organic alcohols.

Level

Post-16.

Timing

20 min.

Description

These two important alcohols can be chemically distinguished by using the iodoform reaction. Students do this reaction at microscale using a well-plate, one advantage being that no heating is required.

$$C_2H_5OH \xrightarrow{I_2} CH_3CHO \xrightarrow{I_2} CI_3CHO \xrightarrow{NaOH} CHI_3 \text{ (s)} + HCOONa$$

Apparatus (per group)

▼ One plastic well-plate (24 well size, *eg* Sigma ref: M 9655).

Chemicals (per group)

Solutions contained in plastic pipettes, see p. 2

▼ Methanol

▼ Ethanol

▼ Iodine solution (0.05 mol dm^{-3}) dissolved in potassium iodide (0.2 mol dm^{-3})

▼ Sodium hydroxide   1 mol dm^{-3}.

Note

Both methanol and ethanol (and other alcohols and organic liquids) can be stored in plastic pipettes and easily dispensed. The main difficulty for students arises from the fact that if the pipettes are squeezed too hard the alcohols come out of the pipette in a stream (because of their low surface tension). Students must handle the pipettes very carefully and some practice is required before proceeding with this experiment.

Observations

After a short time a cloudiness is seen in the ethanol well while the methanol well remains clear. After a few minutes a yellow precipitate forms which has a distinctive antiseptic smell. Compounds that contain a methyl group adjacent to a carbonyl group (methyl ketone) give a positive result with this test, as do secondary alcohols which can be oxidised to ketones.

Safety

Students must wear eye protection.
It is the responsibility of the teacher to carry out a risk assessment.

46. The formation of solid derivatives of aldehydes and ketones using 2,4-dinitrophenylhydrazine (Brady's Test)

Topic

Aldehydes and ketones.

Level

Post-16.

Timing

15 min.

Description

In this experiment various aldehydes and ketones are mixed with 2,4-dinitrophenylhydrazine solution in a well-plate to form solid derivatives. Two alcohols are also included to show that they do not react.

Apparatus (per group)

▼   One plastic well-plate (24 well size, *eg* Sigma ref: M 9655).

Chemicals (per group)

Solutions contained in plastic pipettes, see p. 2

▼   Ethanol

▼   Propanone

▼   *p*-Methoxybenzaldehyde (or other aromatic aldehyde or ketone)

▼   Methanol

▼   Ethanal

▼   Solution of 2,4-dinitrophenylhydrazine (prepared as described below) which requires 24 hours to dissolve completely.

Preparing a solution of dinitrophenylhydrazine

1.   Weigh out 0.5 g of 2,4-dinitrophenylhydrazine and dissolve in 10 cm^3 of concentrated sulphuric acid in a 100 cm^3 beaker. (Leave overnight to allow the compound to dissolve completely.)

2.   Add this solution carefully to a solution of 15 cm^3 of ethanol and 5 cm^3 of deionised water. This gives the yellow-brown stock solution.

The solution can be used for about 10 days but it gradually deteriorates as a precipitate forms.

This procedure is based on that described in L. Harwood and R. Moody, *Experimental organic chemistry – principles and practice*, p 242. Oxford: Blackwell Scientific, 1989.

THE ROYAL
SOCIETY OF
CHEMISTRY

Observations

Compound	Observation
Ethanal	Immediate yellow precipitate
Propanone	Yellow, crystalline precipitate forms slowly (after 1-2 minutes)
p-Methoxybenzaldehyde (or other aromatic aldehyde or ketone)	Immediate red precipitate
Methanol	No change
Ethanol	No change

The reaction between propanone and dinitrophenylhydrazine

$$\underset{H_3C}{\overset{H_3C}{>}}C=O + NH_2-NH-\overset{O_2N}{\underset{}{\bigcirc}}-NO_2$$

$$\downarrow$$

$$\underset{H_3C}{\overset{H_3C}{>}}C=N-NH-\overset{O_2N}{\underset{}{\bigcirc}}-NO_2 + H_2O$$

Safety

Students must wear eye protection.
It is the responsibility of the teacher to carry out a risk assessment.

THE ROYAL
SOCIETY OF
CHEMISTRY

47. Testing for unsaturation using potassium manganate(VII)

Topic

Unsaturation.

Level

Post-16.

Timing

20 min.

Description

A solution of potassium manganate(VII) in propanone is used to detect whether an organic compound is unsaturated. In this experiment the propanone solution is made up and stored in a plastic pipette. The solution mixes easily with non-polar organic compounds such as cyclohexane, cyclohexene and limonene. Unsaturated compounds turn the solution a brownish colour as the manganese(VII) is reduced to manganese(IV) – *ie* MnO_2.

Apparatus (per group)

▼   Plastic pipettes

▼   One plastic petri dish

▼   One 10 cm^3 beaker

▼   Scissors.

Chemicals (per group)

▼   Propanone

▼   Potassium manganate(VII) crystals

▼   Cyclohexane

▼   Cyclohexene

▼   Limonene.

Safety

Students must wear eye protection.
It is the responsibility of the teacher to carry out a risk assessment.

THE ROYAL
SOCIETY OF
CHEMISTRY

48. Preparing and testing ethyne

Topic

Ethyne.

Level

Post-16.

Timing

20 min.

Description

In this experiment ethyne gas is generated inside a plastic petri dish by reacting calcium carbide and water. The gas is tested using a solution of potassium manganate(VII) in propanone, which turns from purple to brown. When the reaction has finished the residue is tested with a drop of full-range indicator solution.

Apparatus (per group)

▼   One clear plastic sheet (*eg* ohp sheet)

▼   Plastic pipettes

▼   One plastic petri dish (5.5 cm diameter, *eg* Philip Harris ref: Y 36340)

▼   One 10 cm³ beaker

▼   Scissors

▼   Tweezers.

Chemicals (per group)

Solutions and liquids contained in plastic pipettes, see p. 2

▼   Propanone

▼   Water

▼   Calcium carbide – small lumps

▼   Potassium manganate(VII) crystals.

Observations

Calcium carbide reacts vigorously with water. The ethyne gas produced turns potassium manganate(VII) from purple to brown indicating ethyne is unsaturated. The residue turns full-range indicator solution purple owing to the presence of the alkaline calcium hydroxide.

This experiment illustrates an interesting link between inorganic and organic chemistry. However, calcium carbide is an expensive material. It is prepared by heating a mixture of calcium oxide and coke to *ca* 2000 ºC in an electric furnace.

Students could also be told about Wohler's experiment in 1828 on heating ammonium cyanate and obtaining urea thus demolishing the vital force theory (that organic compounds could only be obtained from living things).

Safety

Students must wear eye protection.

It is the responsibility of the teacher to carry out a risk assessment.

THE ROYAL
SOCIETY OF
CHEMISTRY

49. Testing for unsaturation using bromine

Topic

Reactions of bromine and unsaturated organic molecules.

Level

Post-16.

Timing

15 min.

Description

A solution of bromine in hexane is used to detect whether an organic compound is unsaturated. In this experiment the hexane solution is made as described below. The solution mixes easily with non-polar organic compounds such as cyclohexane cyclohexene and limonene. Unsaturated compounds rapidly decolorise the bromine.

Apparatus (per group)

▼   Plastic pipettes

▼   One plastic petri dish.

Chemicals (per group)

Chemicals contained in plastic pipettes, p. 2

▼   Cyclohexane

▼   Cyclohexene

▼   Limonene

▼   Solution of bromine in hexane (see below).

Preparing a solution of bromine in hexane

In this experiment elemental bromine is formed by a reverse disproportionation reaction between bromate, bromide and acid:

$$BrO_3^-(aq) + 5Br^-(aq) + 6H^+(aq) \rightarrow 3Br_2(aq) + 3H_2O(l)$$

The bromine is then extracted into hexane in a plastic pipette which serves as a separating funnel. The resulting solution is decanted into a well-plate chamber and can then be used to test for unsaturation in organic compounds.

One particular safety advantage of this microscale technique is that it eliminates the need for bottles of bromine and its associated hazards. In addition, the experiment is quite fun to do and illustrates several important chemical principles such as disproportionation, extraction techniques and hydrophilic/hydrophobic properties.

THE ROYAL
SOCIETY OF
CHEMISTRY

Apparatus (per group)

▼   One plastic well-plate (24 well size, *eg* Sigma ref: M9655)

▼   One plastic pipette (standard form, *eg* Aldrich ref: 213, 500-3).

Chemicals (per group)

Solutions and liquids contained in plastic pipettes, see p. 2

▼   Potassium bromate(V) $0.1 \ mol \ dm^{-3}$

▼   Potassium bromide $0.2 \ mol \ dm^{-3}$

▼   Hydrochloric acid $1 \ mol \ dm^{-3}$

▼   Hexane.

Safety

Students must wear eye protection.
     It is the responsibility of the teacher to carry out a risk assessment.

THE ROYAL
SOCIETY OF
CHEMISTRY

50. The oxidation of alcohols

Topic

Organic chemistry – alcohols.

Level

Post-16.

Timing

20 min.

Description

In this experiment students look for colour changes when drops of various alcohols are added to acidified potassium dichromate solution in a well-plate. The experiment has advantages over traditional methods in that only very small amounts of each alcohol are needed and no heating is required.

Apparatus (per group)

▼   Well-plate (24 wells) – eg Sigma ref: M9655.

Chemicals (per group)

▼   Methanol

▼   Ethanol

▼   Propan-1-ol

▼   Propan-2-ol

▼   2-Methylpropan-2-ol (or other tertiary alcohol)

Solution contained in plastic pipette, see p. 2

▼   Acidified potassium dichromate.

(To prepare a stock solution: dissolve 2 g of potassium dichromate in 80 cm³ of deionised water and carefully add 10 cm³ of concentrated sulphuric acid).

Observations

With the primary alcohols (methanol, ethanol and propan-1-ol) the dichromate solution starts to turn green after a few minutes indicating that oxidation is taking place although the methanol usually seems to be slower than the other two. With propan-2-ol (secondary alcohol) the colour change occurs, but is rather slow.

With 2-methylpropan-2-ol no oxidation occurs and the dichromate solution does not change colour.

Safety

Students must wear eye protection.

It is the responsibility of the teacher to carry out a risk assessment.

THE ROYAL
SOCIETY OF
CHEMISTRY

51. The oxidation of cyclohexanol by potassium dichromate

Topic

Organic oxidation reactions, reactions of dichromate.

Level

Post-16.

Timing

15 min.

Description

In this experiment cyclohexanol is oxidised to cyclohexanone using an acidified solution of potassium dichromate. A drop of cyclohexanol is added to a pool of the dichromate and over the next few minutes a great deal of surface activity is observed. This is due to the differences in physical properties such as surface tension and viscosity of the product (cyclohexanone) compared to the reactant (cyclohexanol).

Apparatus (per group)

▼　　One clear plastic sheet (*eg* ohp sheet)

▼　　One plastic pipette (standard form, *eg* Aldrich ref: Z13, 500-3)

▼　　One plastic petri dish (5.5 cm size).

Chemicals (per group)

▼　　Cyclohexanol

(Note: Cyclohexanol is an interesting substance as its melting point is 20–22 °C – *ca* room temperature. If this experiment is done on a warm day the cyclohexanol is liquid and therefore easier to sample than on a cold day when it is solid.)

▼　　Acidified potassium dichromate solution

Dissolve 2 g of potassium dichromate in 80 cm^3 of deionised water and add 10cm^3 of concentrated sulphuric acid to the solution.

Safety

Students must wear eye protection.
    It is the responsibility of the teacher to a carry out a risk assessment.

THE ROYAL
SOCIETY OF
CHEMISTRY

52. The oxidation of cyclohexanol by nitric acid

Topic

Alcohols, carboxylic acids, oxidations.

Level

Post-16.

Timing

20 min.

Description

In this experiment students convert cyclohexanol to 1,6-hexanedioic acid (adipic acid) using a ring opening oxidation with nitric acid. Since 1,6-hexanedioic acid is a solid a melting point measurement can be done on the product (mp 152 °C).

Apparatus (per group)

▼   One 100 cm³ beaker

▼   Hot plate

▼   Three plastic pipettes

▼   One 50 cm³ beaker

▼   One test-tube.

Chemicals (per group)

▼   Cyclohexanol

▼   Nitric acid (*ca* 5 mol dm⁻³, concentrated nitric acid: deionised water 1:1).

Observations

A white crystalline solid should slowly form when the test-tube is cooled in the ice bath. The solid might be slightly brown in colour due to impurities when first filtered off, but this discoloration is removed by washing with water.

Reference

S. Breuer, *Microscale practical organic chemistry*, expt 26. Lancaster: Lancaster University, 1991.

Safety

Students must wear eye protection. The reaction should be done in a fume cupboard. It is the responsibility of the teacher to carry out a risk assessment.

THE ROYAL
SOCIETY OF
CHEMISTRY

53. The microscale synthesis of aspirin

Topic

Medicines and organic synthesis.

Level

Post-16.

Timing

20 min.

Description

In this experiment students do a microscale esterification reaction between 2-hydroxybenzoic acid (salicylic acid) and ethanoic anhydride using phosphoric acid as a catalyst.

Apparatus (per group)

- ▼ One 10 cm^3 beaker
- ▼ Hot plate
- ▼ One 5 cm^3 measuring cylinder
- ▼ One 50 cm^3 beaker
- ▼ One test-tube
- ▼ Small filter funnel.

Chemicals (per group)

- ▼ 2-Hydroxybenzoic acid (salicylic acid)
- ▼ Ethanoic anhydride
- ▼ Phosphoric acid (85%).

Observations

This esterification reaction, which uses reactive ethanoic anhydride and phosphoric acid catalyst, is quite fast at microscale. A good yield of white crystals should be formed.

Reference

J.Chem.Ed., 1987, **64**, 440.

Safety

Students must wear eye protection. This experiment should be done in a fume cupboard.
It is the responsibility of the teacher to carry out a risk assessment.

THE ROYAL
SOCIETY OF
CHEMISTRY

54. The analysis of aspirin tablets

Topic

Organic chemistry/chemical analysis.

Level

Post-16.

Timing

20 min.

Description

In this experiment students measure the amount of free 2-hydroxybenzoic acid (salicylic acid) in 2-ethanoyloxybenzenecarboxylic acid (aspirin) tablets. 2-hydroxybenzoic acid (salicylic acid), being a substituted phenol, reacts with Fe^{3+} ions to produce a purple colour. The colour is matched against that produced by a set of standard solutions of 2-hydroxybenzoic acid (salicylic acid) in a well-plate.

Apparatus (per group)

▼   One 24-well plate

▼   One 100 cm^3 beaker

▼   Cotton wool

▼   One plastic pipette (standard form, *eg* Aldrich ref: Z13, 500-3)

▼   Two plastic pipettes (fine tip, *eg* Aldrich ref: Z13, 503-8).

▼   Sheet for microscale filtration technique.

Chemicals (per group)

▼   Various 2-ethanoyloxybenzenecarboxylic acid (aspirin) tablets

Solutions contained in plastic pipettes (fine-tip), see p. 2

▼   Iron(III) nitrate solution

▼   2-Hydroxybenzoic acid (salicylic acid) (working) solution

▼   Deionised water.

1.   Stock 2-hydroxybenzoic acid (salicylic acid) solution (0.1% w/v)

       Dissolve 0.100 g of 2-hydroxybenzoic acid (salicylic acid) in *ca* 20 cm^3 of a 1:1 mixture of ethanol and deionised water in a 100 cm^3 beaker. Make up to 100 cm^3 in a volumetric flask.

2.   Working 2-hydroxybenzoic acid (salicylic acid) solution (0.0025 g 2-hydroxybenzoic acid (salicylic acid) /25 cm^3)

       Dilute 2.5 cm^3 of the stock solution to 25 cm^3 in a volumetric flask with a 1:1 ethanol/water mixture.

3.   Iron(III) nitrate solution, 0.1 mol dm^{-3}.

Observations

The set of standard solutions should give a range of intensities of a bluish colour. Students should be careful to add the correct number of drops as indicated. The experiment works best with old tablets containing some free 2-hydroxybenzoic acid (salicylic acid). New tablets with minimal free acid do not give any blue coloration but merely the colour of iron(III) in solution (yellow) so they do not fit into the range of standards.

The equation by which 2-hydroxybenzoic acid (salicylic acid) is formed is:

2-Ethanoyloxybenzenecarboxylic acid
(Aspirin)

2-Hydroxybenzoic acid
(Salicylic acid)

This experiment gives students an opportunity to consider the practical effect of equilibrium. Old 2-ethanoyloxybenzenecarboxylic acid (aspirin) tablets, which may have become damp with time, will contain more free 2-hydroxybenzoic acid (salicylic acid) because the presence of water causes the position of equilibrium to be shifted to the right in the above equation.

Reference

This experiment is based on a similar procedure given in the publication G. Rayner-Canham and A. Slater, *Microscale chemistry – laboratory manual*. Don Mills, Ontario: Addison-Wesley, 1994.

Safety

Students must wear eye protection.

It is the responsibility of the teacher to carry out a risk assessment.

THE ROYAL
SOCIETY OF
CHEMISTRY

55. The conversion of alcohols to halogenoalkanes

Topic

Alcohols, halogenoalkanes, substitution reactions.

Level

Post-16.

Timing

20 min.

Description

This experiment uses phosphorus tribromide to effect the nucleophilic substitution of the hydroxy group in an alcohol by bromine to form a halogenoalkane. The conversion is simpler and quicker to perform than the macroscale method which uses solid potassium bromide and concentrated sulphuric acid followed by distillation. However, it must be done in a fume cupboard because phosphorus tribromide is a hazardous reagent and reacts vigorously with the alcohol. Although teachers may want their students to use the traditional method, this procedure is given as an interesting alternative which could be done either as a teacher demonstration or by students.

Apparatus (per group)

▼   One test-tube

▼   One 5 cm³ measuring cylinder

▼   Glass pipettes.

Chemicals (per group)

▼   Ethanol

▼   Phosphorus tribromide

▼   Deionised water.

Observations

The phosphorus tribromide reacts vigorously with the ethanol and it is important that it is added only one drop at a time. On adding water there may be some reaction between the water and any unreacted phosphorus tribromide. A few globules of bromoethane will be observed at the bottom of the specimen tube.

It is worth pointing out to students that there are parallels with biological processes in which hydroxy groups (on sugars) are first phosphorylated before being substituted or eliminated.

Safety

Students must wear eye protection.
It is the responsibility of the teacher to carry out a risk assessment.

THE ROYAL
SOCIETY OF
CHEMISTRY

56. The microscale synthesis of azo dyes

Topic

Organic chemistry – azo dyes.

Level

Post-16.

Timing

20 min.

Description

In this experiment students prepare an azo dye and use it to dye a piece of cotton.
The synthesis is unusual in that whereas most organic syntheses require ambient or
elevated temperature, this synthesis requires low temperatures.

Apparatus (per group)

▼   Three 10 cm³ beakers

▼   Thermometer

▼   Tweezers.

Chemicals (per group)

▼   Ice

▼   Aminobenzene (aniline)

▼   Hydrochloric acid

▼   Sodium nitrite

▼   2-Naphthol (also called β-Naphthol; naphthalene-2-ol)

▼   Sodium hydroxide solution   2 mol dm⁻³

▼   Ethanol

▼   Urea.

Observations

The orange-red azo dye forms in the fibres of the cotton, dyeing the cloth. The
melting point of 1-phenylazo-2-naphthol is 133 °C.

Note

The urea decomposes excess HNO_2 formed and prevents many side reactions from
occurring. A better 'red' dye is usually produced.

Reference

S. W. Breuer, *Microscale practical organic chemistry*. Lancaster: Lancaster University,
1991.

Safety

Students must wear eye protection. This experiment should be done in a fume
cupboard.
    It is the responsibility of the teacher to carry out a risk assessment.

THE ROYAL
SOCIETY OF
CHEMISTRY

57. The microscale synthesis of indigo dye

Topic

Colour chemistry, dyes. Organic chemistry – dyes, synthesis.

Level

Pre-16 and post-16.

Timing

10 min.

Description

In this experiment students prepare the dye indigo by the condensation of
2-nitrobenzaldehyde with propanone in the presence of sodium hydroxide. The
experiment is very easy to do.

Apparatus (per group)

▼   One 10 cm³ beaker or a test-tube

▼   Small filter funnel and filter paper.

Chemicals (per group)

Solutions contained in plastic pipettes, see p. 2

▼   Sodium hydroxide   0.5 mol dm⁻³

▼   Deionised water

▼   Propanone

▼   2-Nitrobenzaldehyde.

Observations

2-Nitrobenzaldehyde dissolves in propanone to form a pale yellow solution. When
sodium hydroxide solution is added the solution darkens after a few seconds and a
purple precipitate of indigo forms. The synthesis is very simple and quick to perform
yet it is mechanistically complex involving a series of condensations,
disproportionations and oxidations. The sequence of reactions is given in the *Journal
of Chemical Education* and this, together with the structures of the two forms of
indigo, is shown overleaf. There is one unusual step which produces ethanoic acid
and what appears to be a hydroxy-derivative of indoxyl prior to the final oxidation to
form indigo (indigotin).

It is worthwhile pointing out to students that it took Baeyer 14 years (from 1865 to
1879) to find a route for the original synthesis of indigo and even when he did it took
him another four years to deduce the correct formula (in 1883)!

The indigo can be filtered off and then dissolved in an alkaline solution of sodium
dithionite to form the colourless, soluble form (leucoindigo). A piece of cotton cloth
dipped into this solution and then exposed to air produces the indigo-dyed fabric –
this is the procedure used in vat dyeing.

THE ROYAL
SOCIETY OF
CHEMISTRY

Colourless   Blue

The preparation of indigo from 2-nitrobenzaldehyde and propanone

References

1.   *J.Chem.Ed.*, 68, **10**, 1991.

2.   S.W. Breuer, *Microscale practical organic chemistry*. Lancaster: Lancaster
University, 1991.

Safety

Students must wear ear eye protection.
    Students should wear gloves when weighing out the 2-nitrobenzaldehyde.
Alternatively, it could be pre-weighed in the beaker before the lesson, especially for
pre-16 classes.
    It is the responsibility of the teacher to carry out a risk assessment.

THE ROYAL
SOCIETY OF
CHEMISTRY

58. The preparation of ethyl benzoate

Topic

Derivatives of carboxylic acids.

Level

Post-16.

Timing

20 min.

Description

In this experiment students generate ethyl benzoate by warming benzoic acid with ethanol in a plastic pipette in a water bath. The ester is identified by smell.

Apparatus (per group)

▼   One 5 cm³ measuring cylinder

▼   One 10 cm³ beaker

▼   Three plastic pipettes.

Chemicals (per group)

▼   Ethanol

▼   Benzoic acid

▼   Concentrated sulphuric acid.

Observations

The smell of the ester should be apparent. It should be possible to generate other esters by using different combinations of carboxylic acids and alcohols.

References

1.   G. Rayner-Canham and A. Slater, *Microscale chemistry*, pp186 expt 57. Don Mills, Ontario: Addison-Wesley, 1994.

2.   S. W. Breuer, *Microscale practical organic chemistry*, expt 26. Lancaster: Lancaster University, 1991.

Safety

Students must wear eye protection. This experiment must be done in a fume cupboard.
It is the responsibility of the teacher to carry out a risk assessment.

THE ROYAL
SOCIETY OF
CHEMISTRY

59. The treatment of oil spills

Topic

Pollution control. Polymers – uses of intermolecular bonding.

Level

All.

Timing

10 min.

Description

In this experiment, oil or paraffin is added to some water in a beaker to simulate an oil spill. A special powdered polymer is then sprinkled on top. On stirring the polymer absorbs the hydrocarbon molecules and a rubbery solid is formed which can then be scooped up. The experiment is quite fun to do and provides several interesting points for follow-up discussion in both theoretical and applied chemistry (pollution and its control).

Apparatus (per group)

▼   One student worksheet

▼   One 100cm³ beaker

▼   Plastic pipette

▼   Scissors.

Chemicals (per group)

▼   Soil-moist hydrocarbon polymer (see below)

▼   Oil or paraffin.

Observations

On adding the polymer, and stirring, a rubbery solid is formed very quickly and the layer of oil/ paraffin disappears.

Note

The essential ingredient in this experiment is the powdered polymer which can be obtained from Educational Innovations Inc, Cos Cob, Connecticut, USA at a price of $5 per oz. With careful use 1 oz should provide enough for many experiments! The polymer itself is a copolymer of acrylamide and hydroxymethylmethacrylate, crosslinked and dehydrated. A similar substance is produced commercially by BP under the tradename Rigidoil.

Safety

Students must wear eye protection.
    It is the responsibility of the teacher to carry out a risk assessment.

THE ROYAL
SOCIETY OF
CHEMISTRY

60. Detecting starch in food

Topic

Tests for substances in food.

Level

Pre-16.

Timing

10 min.

Description

In this experiment students test for the presence of starch in various foodstuffs –
eg cereals, bread and rice.

Apparatus (per group)

▼   One student worksheet

▼   One clear plastic sheet (*eg* ohp sheet)

▼   Tweezers.

Chemicals and foodstuffs (per group)

▼   Bleach (contained in a plastic pipette)

▼   Potassium iodide crystals

▼   Various foodstuffs containing starch (*eg* cereals and bread) and some not
     containing starch (*eg* mushrooms).

Observations

The characteristic blue-black colour of the starch-iodine complex is seen almost
immediately the drop of bleach touches the potassium iodide crystal on the foodstuff.

Reference

Edward L. Waterman, *Workshop on small-scale chemistry*. NSTA Annual
Convention, Philadelphia, US, March 1995.

Safety

Students must wear eye protection.
   It is the responsibility of the teacher to carry out a risk assessment.